ABOUT STARS

Their Formation, Evolution, Compositions, Locations and Companions

Other World Scientific Titles by the Author

Colour: How We See It and How We Use It
ISBN: 978-1-78634-084-9
ISBN: 978-1-78634-085-6 (pbk)

Everyday Probability and Statistics: Health, Elections, Gambling and War
Second Edition
ISBN: 978-1-84816-761-2
ISBN: 978-1-84816-762-9 (pbk)

The Formation of the Solar System: Theories Old and New
Second Edition
ISBN: 978-1-78326-521-3
ISBN: 978-1-78326-522-0 (pbk)

The Fundamentals of Imaging: From Particles to Galaxies
ISBN: 978-1-84816-684-4
ISBN: 978-1-84816-685-1 (pbk)

Materials, Matter and Particles: A Brief History
ISBN: 978-1-84816-459-8
ISBN: 978-1-84816-460-4 (pbk)

On the Origin of Planets: By Means of Natural Simple Processes
ISBN: 978-1-84816-598-4
ISBN: 978-1-84816-599-1 (pbk)

Resonance: Applications in Physical Science
ISBN: 978-1-78326-538-1
ISBN: 978-1-78326-539-8 (pbk)

Time, Space, Stars and Man: The Story of the Big Bang
Second Edition
ISBN: 978-1-84816-933-3
ISBN: 978-1-84816-934-0 (pbk)

ABOUT STARS

Their Formation, Evolution, Compositions, Locations and Companions

Michael M. Woolfson
University of York, UK

NEW JERSEY • LONDON • SINGAPORE • BEIJING • SHANGHAI • HONG KONG • TAIPEI • CHENNAI • TOKYO

Published by

World Scientific Publishing Europe Ltd.

57 Shelton Street, Covent Garden, London WC2H 9HE

Head office: 5 Toh Tuck Link, Singapore 596224

USA office: 27 Warren Street, Suite 401-402, Hackensack, NJ 07601

Library of Congress Cataloging-in-Publication Data

Names: Woolfson, Michael M. (Michael Mark), author.

Title: About stars : their formation, evolution, compositions, locations and companions / Michael Woolfson.

Description: Singapore ; Hackensack, NJ World Scientific Publishing Co. Pte. Ltd., [2019] | Includes bibliographical references and index.

Identifiers: LCCN 2019020006 | ISBN 9781786347121 | ISBN 1786347121 (hardcover) | ISBN 9781786347251 (paperback) | ISBN 1786347253 (paperback)

Subjects: LCSH: Stars. | Stars--Formation. | Stars--Evolution.

Classification: LCC QB801 .W69 2019 | DDC 523.8--dc23

LC record available at https://lccn.loc.gov/2019020006

British Library Cataloguing-in-Publication Data

A catalogue record for this book is available from the British Library.

For any available supplementary material, please visit https://www.worldscientific.com/worldscibooks/10.1142/Q0211#t=suppl

Desk Editors: Herbert Moses/Jennifer Brough/Shi Ying Koe

Typeset by Stallion Press
Email: enquiries@stallionpress.com

Printed in Singapore

About the Author

Courtesy:
Alan Culpan

Michael M. Woolfson is Professor Emeritus in Theoretical Physics at the University of York. His main fields of research are the development of methods of solving crystal structures, particularly proteins and in the study of star and planet formation. He has published 25 books on various scientific topics.

Contents

Introduction

If we look up at the sky on a cloudless night, well away from light pollution, we see a dense distribution of points of light. Even with a modest telescope the number of these light sources that can be seen greatly increases. We know what these points of light are; they are mostly stars, many like our closest star, the Sun, but some very different. They are so far away that by any physical test we can apply they are indistinguishable from point sources. However, astronomers want to find the physical properties of these stars, or other objects if that is what they are — their sizes, masses, temperatures and compositions — and how distant they are. To do this they have used our knowledge of physics that has been built up over the years by both experimentalists and theoreticians.

When we find the properties of stars they sometimes reveal that their matter is in states that we cannot reproduce on Earth. For example, a type of star known as a white dwarf has a density such that a teaspoonful of its material has a mass of several tonnes. This provides a test for our theories and it turns out that the field of quantum mechanics can provide a theoretical explanation, not only for the structure of white dwarfs but also of an even more exotic body, a neutron star where a teaspoonful of its matter has a mass of a billion tonnes.

The first part of this book concentrates on the techniques used to find their externally observable properties and locations of stars but does also deal with the internal structures of the most common types of star. It is not an exhaustive treatise on stars but provides sufficient knowledge to give a basis for further study. It begins with a description of the process that provided the material for the production of stars and everything else in the Universe, including us — the Big Bang.

In finding a title for this book the flexibility of the English language was exploited, where there can be several words that are synonymous but also several meanings for one word. So far 'about stars', as described above, has been taken with the meaning of 'concerning stars' but we also deal with exoplanets, the planets 'about stars' in the sense of being 'around stars'. This is a topic, recent by astronomical standards, which is of increasing interest. A range of general properties of exoplanets is described as are two theories, very different in their approach, for how planets may form. Finally there is an account of how various features of the Solar System may have arisen, based on scenarios that depend on one of the theories of planet formation.

Part 1

Creating Material for the First Stars

Chapter 1

The Creation of the First Matter

1.1. The Nature of Matter

Before embarking on the task of explaining what stars are made of, and how their compositions are determined, first the nature of matter will be described and then the way that matter came into existence; as will be explained later, stars play an important role in that story. We start by examining a very common substance with which everyone is familiar — water. A simple experiment, as illustrated in Figure 1.1, breaks down water into its components and illustrates the principle that the myriad of materials with which we are surrounded — wood, steel, plastics and even ourselves — are composed of a finite number of basic components. An electric current is passed through the water by connecting a battery to two electrodes inserted in separate tubes, both initially full of water, and the gasses released at the positive terminal (anode) and at the negative terminal (cathode) are separately collected. This process, called *electrolysis*, breaks down the water into its two basic components — hydrogen, collected over the cathode and oxygen collected over the anode. Here we have a simple example of a *molecule*, water (chemical symbol H_2O), being split up into its elemental components, hydrogen (chemical symbol H) and oxygen (chemical symbol O).

A molecule of water is illustrated schematically in Figure 1.2(a). The two hydrogen atoms are connected to the oxygen atom by *chemical bonds*. A more complex, but quite common, molecule is

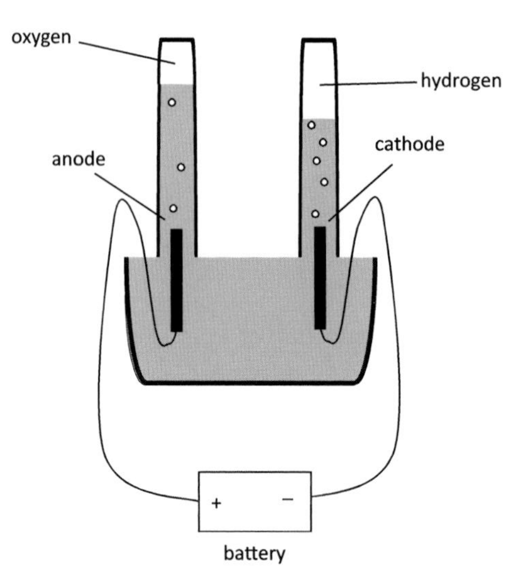

Figure 1.1. The electrolysis of water.

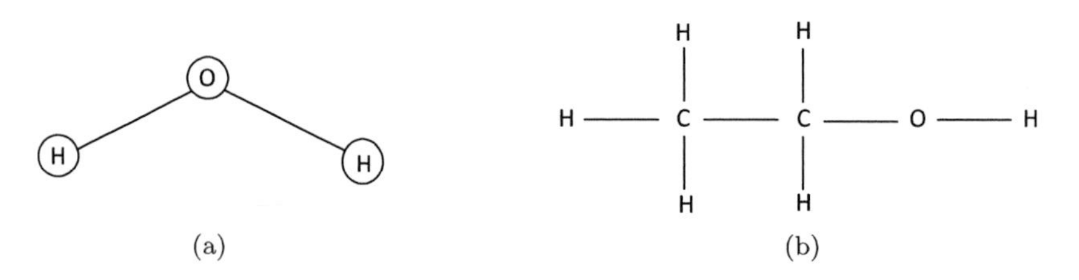

Figure 1.2. Molecules of (a) water and (b) ethyl alcohol.

ethyl alcohol, the essential component of all alcoholic drinks, written in the form C_2H_5OH, where C represents a carbon atom. This is illustrated in Figure 1.2(b). The most complex materials, proteins and viruses containing many thousands of atoms, can be similarly represented, as connected atoms of a finite number of types.

Before nuclear reactors were available, in which new elements can be created, there were 92 types of atom, *elements*, which occur naturally. Some of them are very familiar, such as carbon and iron, but others, like ytterbium, a silvery-metallic element, and selenium, a non-metallic element related chemically to sulphur, are less familiar to most people. One of the 92 elements, astatine, is so rare that only

(a) (b)

Figure 1.3. The atom men (a) Democritus and (b) John Dalton.

about 30 grams of it exist on Earth at any time. It is produced by the radioactive decay of uranium and thorium but it is radioactive and decays itself, disappearing within a few days.

The idea of an atom originated with the Greek philosopher Democritus (460–357 BC; Figure 1.3(a)), who wondered what would happen if matter were repeatedly divided over and over again. He concluded that eventually one would arrive at an indivisible, indestructible particle of matter. The word *atom* comes from the Greek, *atomos*, which means indivisible. However, in the view of Democritus it was possible to have an atom of water or an atom of wood. The idea of atoms, as we understand them today, comes from the work of the English scientist, John Dalton (1766–1844: Figure 1.3(b)). The way that Dalton was led to the modern idea of atoms was based on the following type of argument. When tin combines with oxygen two different compositions by mass are possible, one with 88.1% tin and the other with 78.7% tin. The masses of oxygen that combine with 100 gm of tin are 13.5 gm and 27.1 gm, closely in the ratio 1:2. Similarly the masses of tin that would combine with 100 gm of oxygen are 740 gm and 369 gm, closely in the ratio 2:1. This could be interpreted as both the tin and the oxygen existing in the form of atoms with a particular ratio of masses of tin and oxygen. By this, and other similar examples, Dalton established the idea of atoms corresponding to chemical elements and of linking atoms to form molecules but, like Democritus, he considered atoms as the ultimate units of matter. Now it is known that atoms can be divided and that they have a sub-structure of even smaller particles.

1.2. The Components of Atoms

While Dalton developed all the basic ideas of modern atomic theory — the idea of atoms corresponding to different elements and how they linked together to form compounds — there was no model for the composition of an atom. Was it a small homogeneous blob of material and, if so, what was its extent and why could it not be subdivided? It took more than fifty years after the death of Dalton for scientists to begin to answer these questions.

1.2.1. *The Discovery and Properties of the Electron*

The first indication that atoms had a sub-structure was given in 1897 by the work of the English physicist Joseph John Thomson (1856–1940; Figure 1.4(a); Nobel Prize in Physics, 1906). He was carrying out experiments, which many others had done previously, of passing a current through gasses at very low pressure contained in a glass tube with an electrode at each end. It was already known that some kind of radiation came from the cathode — radiation called *cathode rays* — but the nature of those rays was unknown. Thomson found that a fine beam of cathode rays was deflected by both electric and magnetic fields. The cathode rays were passed through a combination of a magnetic field and an electric field, using the equipment shown in Figure 1.5. The cathode, C, was at a high negative potential relative to the earthed anode, S_1, so electrons, that we now know were being emitted, were strongly repelled by the cathode towards the anode. Since S_1 and S_2 had narrow slits at right angles to each other, a fine beam of cathode rays was produced.

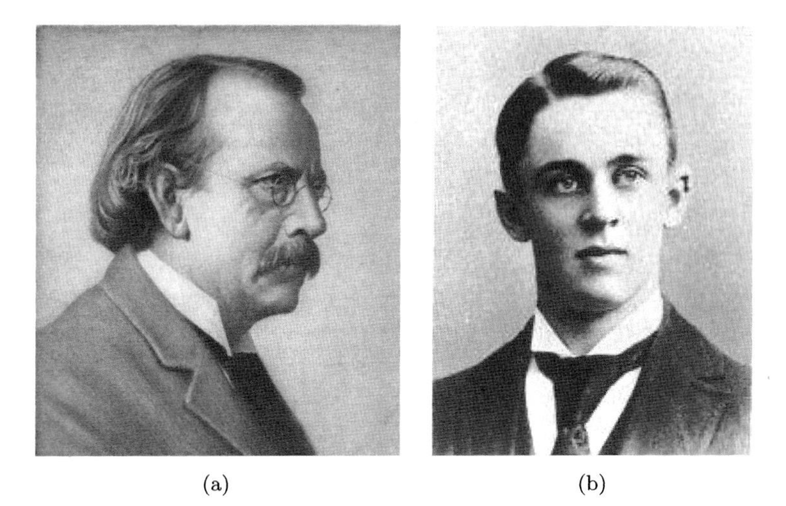

Figure 1.4. The electron men (a) J.J. Thomson (b) R.A. Millikan.

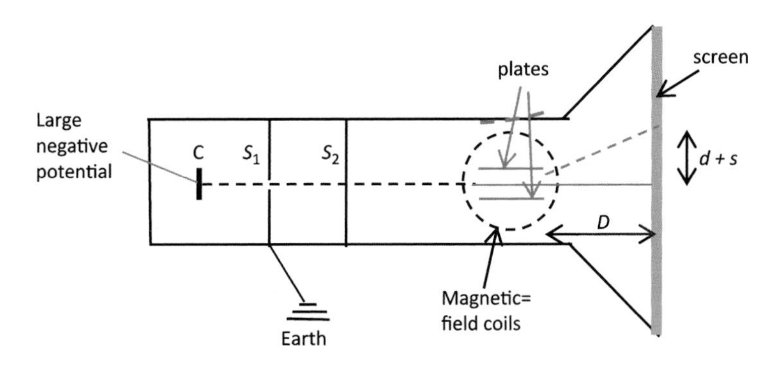

Figure 1.5. A schematic representation of Thomson's cathode-ray experiment.

Thomson's experiment was designed to show that cathode rays consisted of a stream of charged particles of some kind, which he called *corpuscles*, having a charge e and a mass m. The electric field, E, applied a force eE and the deflection it produces depends on the mass of the particle, the deflection decreasing with increasing particle mass. The force due to a magnetic field B is eBv where v is the component of the velocity of the particle perpendicular to the direction of the field. The electric and magnetic fields were arranged to oppose each other to give zero deflection, and the velocity of the cathode rays could then be determined from the ratio E/B.[1] The method of finding e/m will now be described (Thomson, 1906) with the axis of the instrument taken as the x-direction and the direction of deflection of the electrons as the y-direction.

The plates giving the electric field gave the force Ee over a distance L, traversed by the electrons in a time

$$t = \frac{L}{v} = \frac{LB}{E}. \tag{1.1}$$

The acceleration, a, normal to the plates is given by

$$a = \frac{Ee}{m}, \tag{1.2}$$

and after time t the speed in the y-direction is

$$V = at = LB\frac{e}{m}, \tag{1.3}$$

and the beam of electrons would have moved in the y-direction a distance

$$s = \frac{1}{2}ai^2 = \frac{L^2B^2}{2E}\left(\frac{e}{m}\right). \tag{1.4}$$

The electrons thereafter move at a constant speed with component v along x and V along y. If they move a distance D from the edge of the plates to the fluorescent screen, where they are detected then the total deflection is

$$d = D\frac{V}{v} + s = \frac{LB^2}{E}\left(D + \frac{L}{2}\right)\frac{e}{m}, \tag{1.5a}$$

or

$$\frac{e}{m} = \frac{dE}{LB^2}\left(D + \frac{L}{2}\right)^{-1}. \tag{1.5b}$$

[1] In basic SI units the electric field E (volts per metre) has dimensions kg m s^{-3} A^{-1} and the magnetic field has dimensions kg s^{-2} A^{-1}. Hence E/B has dimensions m s^{-1}, i.e. the dimensions of speed.

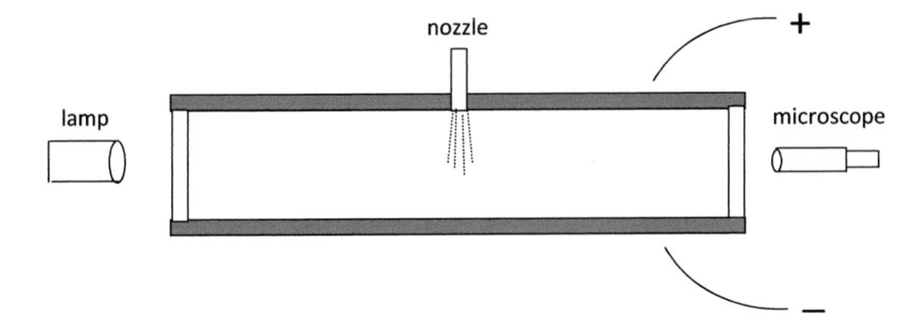

Figure 1.6. A schematic representation of Millikan's oil-drop apparatus.

With d measured and everything else on the right-hand side known, the value of e/m can be found.

The individual values of e and m were found experimentally in 1910 by the American physicist Robert Millikan (1868–1953; Figure 1.4(b): Nobel Prize in Physics, 1923). He suspended oil drops, produced by passing the oil through a spray and charged by exposing them to X-rays, in a vertical electric field, the strength of which was adjusted until a particular drop was stationary, neither rising nor falling (Figure 1.6). The upward force due to the electric field was then equal to the downward force due to the weight of the oil drop. The oil drop was then allowed to fall freely and reached a terminal speed v_T, which was measured. The force on the spherical drop due to Stoke's law then equalled to the weight of the drop, i.e.

$$6\pi\eta a v_T = \frac{4}{3}\eta\rho a^3 g, \tag{1.6}$$

where η is the known viscosity of air, a the radius of the drop, ρ the density of the oil and g the acceleration due to gravity. This gave the radius of the drop and hence its weight. Equating this to Ee when the drop was stationary gave e. The oil-drop charges were always negative and equal to a multiple of a small charge, which Millikan took to be the electron charge, e. With e known, then from Thomson's determination of e/m, the mass of the electron could also be determined. The accepted modern values are $e/m = 1.759 \times 10^{11}$ C kg^{-1} with $e = 1.602 \times 10^{-19}$ coulombs (C)

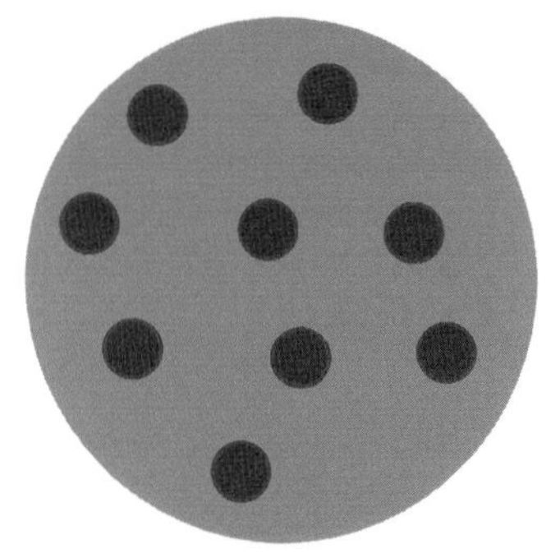

Figure 1.7. Thomson's plum-pudding model of an atom. Negatively charged corpuscles (electrons) are distributed within a spherical cloud of positive charge to give an electrically neutral atom.

and $m = 9.109 \times 10^{-31}$ kg. This mass is 1/1,838 of the mass of the lightest atom, hydrogen.

Atoms are electrically neutral so if electrons, in the form of cathode rays, are coming from atoms then what is left must have a positive charge — it is a positively-charged *ion*. At the beginning of the 20th century experiments were being carried out, similar to those done by Thomson, to measure e/m for various ions and to determine their masses, which turned out to be thousands of times that of the electron. However, the way that the positively-charged ions and the negatively-charged electrons combined to form a neutral atom was still unknown. In 1897 Thomson had proposed a 'plum-pudding' model of an atom in which electrons were situated within a blob of positive charge, like currants in a plum pudding (Figure 1.7).

1.2.2. *The Discovery of Protons and Neutrons*

In 1899 the New Zealander, Ernest Rutherford (1876–1937; Figure 1.8(a); Nobel Prize in Chemistry, 1908), who was working in Manchester, showed that so-called *α-rays* coming from radium were actually particles, now called *α-particles*, with a mass four times that of hydrogen and a positive charge equal in magnitude to that of two electrons (Rutherford, 1908). In 1907, he suggested to his two assistants, Hans Geiger and Ernest Marsden, that they should carry out an experiment in which a stream of α-particles, derived

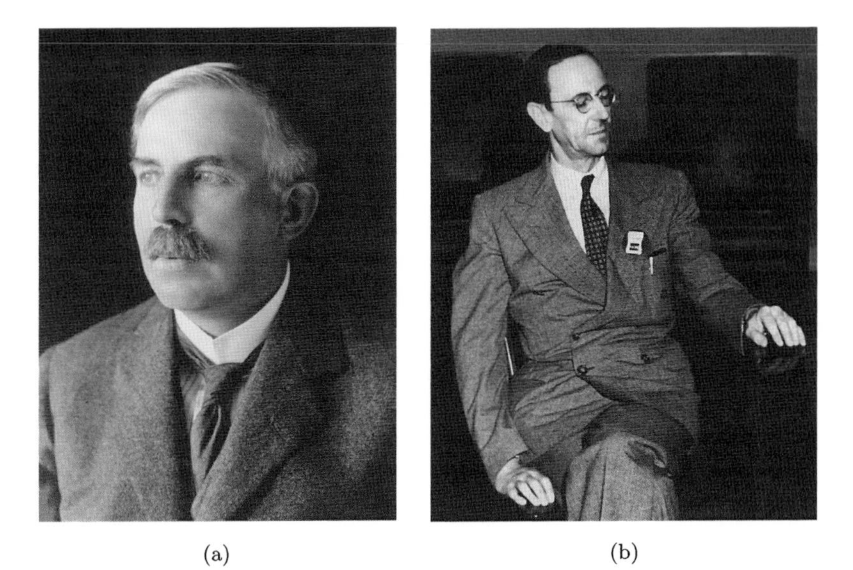

(a) (b)

Figure 1.8. The nucleus men (a) E. Rutherford (b) J. Chadwick.

from a radioactive source, was directed at a thin gold foil and the distribution of scattered α-particles recorded. The experiments were carried out between 1908 and 1913 and gave astonishing results. On the basis of the plum-pudding model there should have been very little scattering of the α-particles — and, indeed, this was true for the great majority of them — but in the experiments a small number were scattered though large angles with some apparently bouncing backwards from the foil. Rutherford described it as being like shells from a naval gun bouncing back from a sheet of tissue paper. This seminal experiment led to a new, and correct, model of an atom. The positive charge is concentrated in a tiny volume — the *atomic nucleus* — and the electrons surround the nucleus in a comparatively large volume so that the bulk of the atom was empty space through which the α-particles were moving. Mostly the positively-charged α-particles were distant from the positively-charged nucleus and so were repelled and scattered through a small angle. However, a small proportion of them passed close to the nucleus and were strongly repelled and scattered through large angles, even to the extent of being scattered backwards.

Since the positive charges of ions were an integral times that of the magnitude of the electronic charge it was clear that there were entities within the nucleus with unit positive charge. In addition the mass of nuclei were integral times a unit mass but the problem was that, if there were just one kind of particle in the nucleus, it would be expected that the number of charge units would be equal to the number of mass units, which was not so. With the exception of hydrogen there were always more units of mass than there were units of charge. One idea to explain this was that the nucleus contained positively-charged particles called *protons* and also some electrons that balanced out part of the positive charge in the nucleus. Later it was postulated that the nucleus contained two types of particle with equal masses, one of which had a positive charge, the *proton*, and the other with no charge, called the *neutron*.

The neutron's existence was confirmed in 1932 by James Chadwick (1894–1974; Figure 1.8(b); Nobel Prize in Physics, 1935), like Rutherford working in Manchester. There is an isotope of a light metal, beryllium, that is radioactive, the emanation from which is electrically neutral and was thought to be electromagnetic radiation. When this emanation fell on various materials their nuclei recoiled and from the energies of the recoiling nuclei Chadwick showed that what was coming from the beryllium were neutral particles with the mass of the proton — in fact, neutrons (Chadwick, 1935).

Now a complete model of an atom was established. There is a very compact nucleus containing protons and neutrons surrounded by electrons, of number equal to that of the protons so that the atom is electrically neutral. The nucleus provides virtually all the mass but the space occupied by electrons defines the size of an atom. To give a picture of the sizes of the nucleus and the whole atom, if the nucleus had the size of a human fist then the radius of the atom would be several kilometres.

Figure 1.9 gives representations of atoms of hydrogen, helium and carbon. The *atomic mass* is the total number of *nucleons* (protons plus neutrons) so it is one for hydrogen, four for helium and twelve for carbon. The *atomic number* is the number of protons in the nucleus — one for hydrogen, two for helium and six for carbon. It is

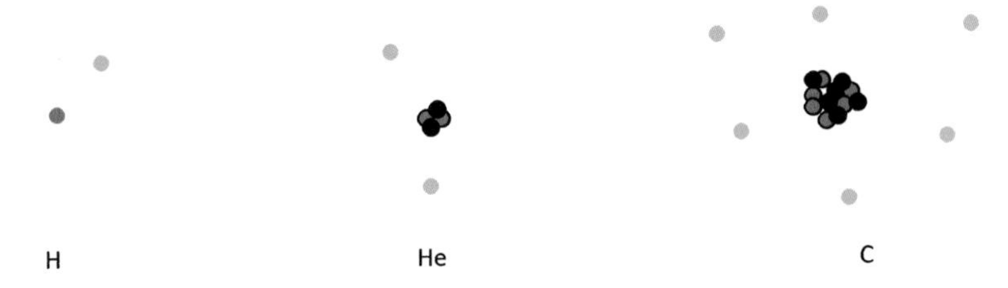

Figure 1.9. Representations of hydrogen (H), helium (He) and carbon (C). Red circles are protons, black circles are neutrons and blue circles are electrons.

the atomic number that defines the element; there is a stable *isotope* of carbon, C-13, with six protons and seven neutrons in the nucleus. Just over one percent of the carbon on Earth (and in you) is C-13. From the previously-given description of the α-particle, it consists of two protons and two neutrons and is the nucleus of a helium atom.

1.2.3. *The Elusive Neutrino*

At this stage the structure of atoms seemed well defined. There are three immutable and fundamental particles — electrons, protons and neutrons — from which all atoms are composed. However, the story of atomic structure had another twist. Some elements have radioactive isotopes that emit *β-particles*, which are very fast-moving electrons. The process that happens in the nucleus is that a neutron converts into a proton plus electron and the very energetic electron is expelled. The nucleus now has an additional proton so corresponds to an element with atomic number one more than the original value. When the electron leaves the nucleus the nucleus recoils, just as a gun recoils when it fires a shell. The direction and speed of the electron and the momentum and energy associated with the recoil can be measured and it was found that neither momentum nor energy was conserved.

In 1930, the Austrian-Swiss (later American) physicist Wolfgang Pauli (1900–1958; Figure 1.10; Nobel Prize in Physics, 1945) proposed that there was an undetected particle, called a *neutrino*, which carried off the missing momentum and energy. It had to be a very strange and elusive particle with very low mass (originally thought to be zero) and very little interaction with matter (Pauli, 1946).

Figure 1.10. Wolfgang Pauli.

Neutrinos are produced by the nuclear reactions that power the Sun and they fall on the Earth with an intensity of 10^{14} neutrinos m^{-2} s^{-1}. They pass through the Sun, the Earth and you with virtually no diminution in numbers. Their detection requires extremely refined and large-scale equipment, built underground to shield it from cosmic rays and holding a large quantity of detecting fluid, to increase the probability that there will be some absorption of neutrinos to detect. Neutrinos may be elusive but of their existence there is now no doubt.

An important property of protons, neutrons and electrons is that they are *fermions*, particles with *half-integral spin,* a description of an intrinsic angular momentum associated with them. The 'half' comes about because when quantum mechanics is applied to the electronic structure of atoms it turns out that all atomic electrons must have an angular momentum associated with their state that is a whole number of the units of $\hbar = h/(2\pi)$, where h is Planck's constant.[2] However, the angular momentum associated with the spins of fermions is $1/2\hbar$ — hence 'half-integral spin'. Associated with their spins, fermions act like tiny magnets. A current moving round a circuit generates a magnetic field and acts like a magnet so it seems natural that the proton and electron, each a charged particle, might give a magnetic field when they possess angular momentum, a property of a spinning body. It is more challenging to consider how a neutron could generate a magnetic field. A conceptual model is that it contains both a positive and balancing negative charge. These spinning in opposite directions would create a magnetic field in the same sense. It is only a conceptual model — not reality!

A property of the neutrino that can be inferred is that it must be a fermion. If a neutron with half-integral spin produced only an electron and proton, each with half-integral spin then spin would not be conserved. The existence of the neutrino solves that problem. Three half-integral spin particles can generate half-integral spin if two of the spins are in an opposite sense.

1.3. Other Particles

Particle physicists do experiments with giant machines that accelerate streams of charged particles up to speeds close to the speed of light and then cause them to collide. The basic design of the largest modern atom-smashing machines is based on that of a *synchrotron*, as shown schematically in Figure 1.11. Particles move on a closed path that they go round repeatedly while being accelerated up to some maximum energy.

The track is an evacuated tube with several straight sections connected by bending magnets that change the direction of the particles so that they go smoothly from one straight section to the next. The straight sections contain devices for accelerating the

[2] Planck's constant occurs frequently in quantum mechanics. Its value is 6.626×10^{-34} m^2 kg s^{-1}.

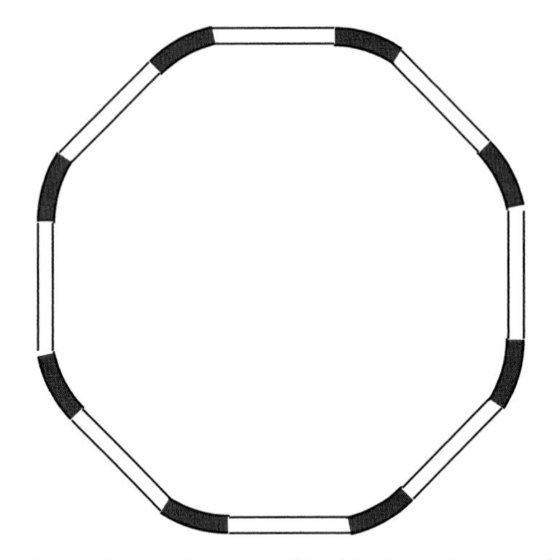

Figure 1.11. A schematic synchrotron. The black sections are bending magnets.

particles, usually in microwave cavities where the charged particles are accelerated by moving with the microwaves much as a surfer rides a wave on the sea. The limit to the energy that can be reached is governed by the fact that when the particles change direction, i.e. are accelerated, in the bending magnets they emit electromagnetic radiation. When the energy of the emitted radiation equals the energy increase in the microwave cavities then the limiting energy is reached.

It is possible to have two beams of particles travelling round the synchrotron in opposite directions in non-intersecting paths. When the required energy is reached their paths can be deflected so as to collide; the resulting particles from the disrupted atoms are recorded by various detectors clustered round the collision region. Figure 1.12 shows part of the ring of the Relativistic Heavy Ion Collider (RHIC) at the Brooklyn National Laboratory, New York, and Figure 1.13 an image showing the tracks of thousands of particles produced by the collision of gold ions.

The *large hadron collider* (LHC), which is 27 km in circumference and spans the Swiss-French border near Geneva, is the most powerful machine of this kind at the present time. It has 9300 liquid-helium-cooled superconducting bending magnets. Counter-rotating beams

Figure 1.12. The relativistic heavy ion collider (Brookhaven National Laboratory).

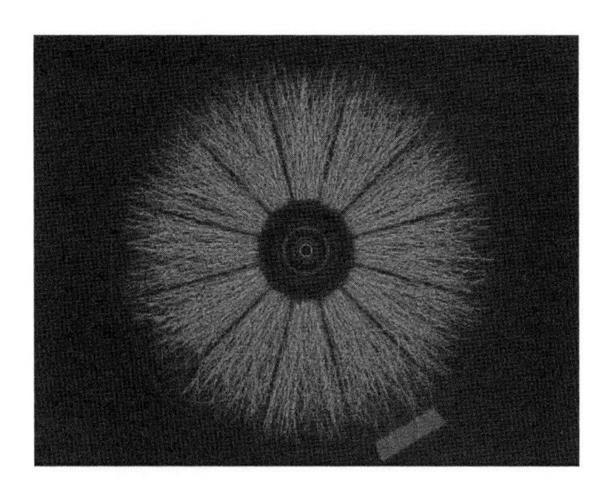

Figure 1.13. Particle tracks from the collision of very high-energy gold ions.

of protons or heavy particles can be produced and then caused to collide when their speeds are 99.999999% of the speed of light. Under such conditions it is hoped that the particles being produced will be similar to those in the early Universe when the energy density was extremely high. It is curious that, in order to observe very tiny and elusive particles, experiments must be on huge scale.

The results of such experiments have revealed a host of new particles. One conclusion from these experiments is that for every particle that exists there is an *antiparticle*, a particle with the same mass and, in the case of a charged particle, with opposite charge. For an electron the antiparticle is the *positron*; there are some radioactive isotopes where a proton in the nucleus converts into a neutron and a positron; the positively charged positron leaves the nucleus at high speed as a β^+-particle. To conserve spin there is also the emission of an *antineutrino*. Other anti-particles that exist are the *antiproton* and the *antineutron*.

If a particle and its antiparticle were to come together their matter would be annihilated with the production of energy in accordance with Albert Einstein's well-known equation linking mass and energy

$$E = mc^2. \tag{1.7}$$

A positron and electron, each with mass 9.101×10^{-31} kg would yield

$$\begin{aligned} E &= 2 \times 9.101 \times 10^{-31} \times (2.998 \times 10^8)^2 \,\mathrm{J} \\ &= 1.64 \times 10^{-13} \mathrm{J} = 1.02 \text{ MeV.}^3 \end{aligned} \tag{1.8}$$

There are circumstances when the reverse process can take place, when energy is converted into matter. When a very high-energy photon (Section 6.2), as exist in cosmic rays, strikes the nucleus of an atmospheric atom then some of it will produce recoil of the nucleus but 1.02 MeV of it can go into *pair production*, the creation of an electron and a positron.

We have now dealt with all the particles that relate to the existence of matter and their associated antiparticles. Before moving on, some of the particles produced by smashing atoms will be described; they will have existed in the early Universe — on the pathway tp producing stars.

---

[3]MeV is one million electron volts. An electron volt is the energy an electron acquires when moving across a potential difference of 1 V. It equals 1.602×10^{-19} J.

Table 1.1. The six leptons.

1	207	3477
electron	muon	tau
electron neutrino	muon neutrino	tau neutrino

1.4. Even More Particles

Particle-physics experiments have revealed many different particles, other than those we have already mentioned that are related to normal matter. Some of these are fundamental particles, in the sense that they have no sub-structure but others are composed of even smaller constituents of matter.

1.4.1. *Leptons*

One of these fundamental particle is the *muon*, a particle similar to the electron in that it has the same charge and is a fermion, but it has a much greater mass and is unstable. It is one of three fundamental particles, in a category known as *leptons* (derived from the Greek word meaning 'thin'), which are related to each other. The electron and muon are two of these and the third is the *tau particle*, again with the same charge as the electron and a fermion but with a much greater mass, almost twice the mass of a nucleon. However, like the muon the tau particle is unstable. For each of these leptons there are neutrinos; the full set of six leptons is shown in Table 1.1 together with their masses in electron units. For each of them there is a corresponding anti-particle.

1.4.2. *Quarks*

In 1961 the American physicist Murray Gell-Mann (b. 1929; Figure 1.14; Nobel Prize in Physics, 1969) developed a theory that postulates the existence of a new set of basic particles called *quarks*, combinations of which give other particles, some directly related to atomic structure and some not (Gell-Mann, 1969). We will concentrate our discussion to how they explain the proton, neutron and their antiparticles.

Figure 1.14. Murray Gell-Mann.

There are six different kinds of quark, defined by their *flavours* as *up*(u), *down*(d), *strange*(s), *charm*(c), *bottom*(b) and *top*(t). The charges associated with quarks are multiples of $1/3$ of an electronic charge, either $-1/3$ or $+2/3$, and for the different quarks the associated fractions of an electronic charge are

$$\mathrm{u}\ 2/3 \quad \mathrm{d}\ -1/3 \quad \mathrm{s}\ -1/3 \quad \mathrm{c}\ 2/3 \quad \mathrm{t}\ 2/3 \quad \mathrm{b}\ -1/3.$$

There are also *antiquarks* to each of these particles with opposite charges — thus

$$\bar{\mathrm{u}}\ -2/3 \quad \bar{\mathrm{d}}\ 1/3 \quad \bar{\mathrm{s}}\ 1/3 \quad \bar{\mathrm{c}}\ -2/3 \quad \bar{\mathrm{t}}\ -2/3 \quad \bar{\mathrm{b}}\ 1/3.$$

In this model, the electron *is* a fundamental particle in its own right but protons and neutrons are formed by combinations of three quarks of the up-down variety. Thus,

u + u + d has a charge, in electron units, $\frac{2}{3} + \frac{2}{3} - \frac{1}{3} = 1$ and is a proton;

$\overline{\mathrm{u}}+\overline{\mathrm{u}}+\overline{\mathrm{d}}$ has a charge, in electron units, $-\frac{2}{3}-\frac{2}{3}+\frac{1}{3}=-1$ and is an antiproton;

$\mathrm{d}+\mathrm{d}+\mathrm{u}$ has a charge, in electron units, $-\frac{1}{3}-\frac{1}{3}+\frac{2}{3}=0$ and is a neutron;

$\overline{\mathrm{d}}+\overline{\mathrm{d}}+\overline{\mathrm{u}}$ has a charge, in electron units, $\frac{1}{3}+\frac{1}{3}-\frac{2}{3}=0$ and is an antineutron.

The existence of quarks has not been experimentally verified but, apart from combinations explaining nucleons, a very large number of experimentally-detected particles can be explained by combining together, either in pairs or in sets of three, combinations from the six quarks. Pairs give particles called mesons, which are *bosons*, i.e. particles with integral spin, and those combining three quarks are *baryons*, which are fermions with half-integral spin and include nucleons. All particles composed of quarks are described as *hadrons*.

1.5. How the Universe Began

It is probably instinctive to believe that the Universe, of which our home, the Earth, is an insignificant part, has always existed, and always will, but that is not the current scientific view. The present standard model takes the view that the matter that comprises the Universe, the space within which that matter exists and the time within which the Universe has evolved, were once non-existent. Such ideas are not easy to comprehend; our beliefs about the nature of time, space and matter are governed by everyday experience that does not encompass the creation of the Universe. We shall now examine the evidence upon which this model is based that makes it, at least, plausible.

1.5.1. *The Expanding Universe*

In Chapter 9, various methods of finding the distances of faraway stars and galaxies and their radial speeds relative to the Milky Way, our galaxy, will be described. The main scientific work of the American astronomer Edwin Hubble (1889–1953; Figure 1.15(a)) was to determine the distances of galaxies. Prior to his work another American astronomer, Vesto Slipher (1875–1969; Figure 1.15(b)),

(a) (b)

Figure 1.15. The Universe men (a) Edwin Hubble (b) Vesto Slipher.

using Doppler-shift observations described in Section 2.10, had measured the radial speeds of distant galaxies and found that all, except some nearby galaxies, were receding from the Milky Way. In 1929, Hubble combined his results with those of Slipher and came up with a remarkable result that distant galaxies were receding from our own galaxy at a speed proportional to their distance. This observation, known as *Hubble's law*, was later reinforced by more precise observations. A plot of radial speed against distance, with distances based on later and better observations than were made by Hubble, is shown in Figure 1.16.

The naïve view that the Universe was eternal and unchanging was inconsistent with Hubble's law that gave clear evidence that the Universe was expanding. There are galaxies for which no distance estimates are available but whose recession speeds can be measured. From Hubble's law their distances can be estimated. The largest recession speed observed is 0.93 times that of light. Since a speed greater than that of light is unobservable according to current theory, it seems that the observed Universe is approaching some theoretical limit.

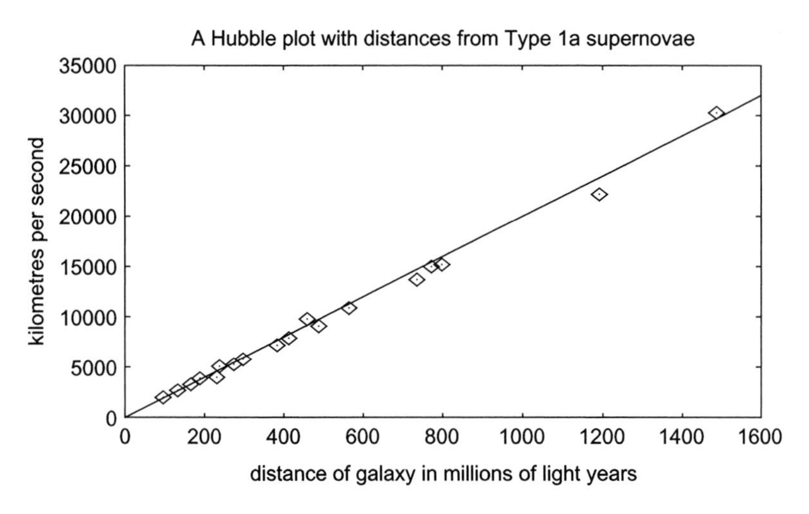

Figure 1.16. An illustration of Hubble's law.

1.5.2. *The Big Bang Hypothesis*

From the Hubble plot it is found that a galaxy at a distance of 3,000 million light years[4] is receding from our galaxy with a speed of 64,000 km s^{-1}. It is not certain that its line of motion passes through our galaxy since there may be a transverse component of motion that we cannot measure for very distant bodies, although they can be measured for nearby stars. However, if we assume that the galaxy's motion was purely radial then we can calculate how long ago — time T_0 prior to now — the galaxy overlapped the Milky Way. This is done by dividing distance by speed. This gives

$$T_0 = \frac{3 \times 10^9 \times 9.461 \times 10^{12}}{6.4 \times 10^3 \times 3.156 \times 10^7} = 14.1 \times 10^9 \quad \text{years},$$

since one year is 3.156×10^7 s. A better estimate is 13.8 billion years. If the same assumption about the radial direction of travel is made for all galaxies then, since speed is proportional to distance, it implies that 13.8 billion years ago all the galaxies occupied the same region of space. Of course, if there were substantial transverse motions then

[4]A light year (ly), the distance light travels in one year, is 9.461×10^{12} km.

Figure 1.17. Monsignor Georges Lemaitre.

the galaxies would not overlap although it would be very likely that they would occupy much less space than now.

The interpretation of the conclusion that galaxies did overlap was very challenging and astronomers wondered what it implied. In 1931, the Belgian astronomer and Roman Catholic priest, Monsignor Georges Lemaitre (1894–1966; Figure 1.17), put forward a theory for the origin of the Universe, which he called the *Primeval Atom Theory* (Lemaitre, 1927) but later became known as the *Big Bang Theory.* The theory proposes that prior to 13.8 billion years ago the Universe, i.e. the matter it contains and the space it occupies did not exist. In fact, time did not exist either so the term just used, 'prior to 13.8 billion years ago' is meaningless — but what else can one say? At that instant all the energy required to create the Universe, concentrated into a point, came into being and from that instant the Universe expanded outwards to give us what we have today. This is difficult, if not impossible, to comprehend and, as previously stated, it implies that matter, space and time came into existence 13.8 billion years ago. Nevertheless, even if we cannot understand the model, it can be expressed in mathematical terms and we can theorize about the very beginnings of time and what was happening in the early Universe.

1.5.3. *The Creation of Particles and Atoms*

High-energy physicists constantly seek to reproduce, as far back as possible, the conditions in the early Universe to discover what kinds of particles existed under those conditions. It is obvious that, however much they try, they will never be able to produce the conditions at the very beginning — there simply is insufficient energy on Earth, even if all the Earth's mass were converted into energy. The best that one can do is to postulate what the early conditions were and what existed under those conditions. It is really a matter of guesswork and intuition. Current ideas about the very early Universe are now described.

From the beginning to 10^{-36} s

During this period, known as the *inflationary period*, the Universe would have expanded extremely rapidly, with the boundary probably moving at greater than the speed of light. The laws of physics that then operated are unknown and would certainly be different from those that apply now. Some energy would have converted into particles of some kind but there would have been little distinction between energy and particles since they would have been converting into each other so rapidly. By the end of this period the physical laws that exist today would be beginning to operate and the Universe would have been a densely-filled mélange of energy and particles.

From 10^{-36} to 10^{-10} s

At the beginning of this period very little matter existed. The way that some of the energy of a photon can be converted into matter by pair production was described in Section 1.3. With photons of extremely high energy, as would have existed during this period, other kinds of pair production would have been possible, in particular the production of quark–antiquark pairs. It would have been a dynamic situation with energy being converted into quark–antiquark pairs but in the reverse direction quarks and antiquarks coming together, with their consequent annihilation producing energy.

The temperature fell as the Universe expanded so that there were fewer high-energy photons available and the rate of quark–antiquark production was reduced. Now something happened for which no explanation is available. By some unknown process the Universe was left with more quarks than antiquarks. This must be true because the Universe contains protons and neutrons made of quarks and not antiprotons and antineutrons made of antiquarks. The excess of quarks over antiquarks was quite small but, after all the antiquarks were removed by combination with quarks, there were sufficient quarks remaining to provide all the matter now in the Universe.

Since we are in the realm of guesswork, and conditions we do not fully understand, we could consider another outlandish possibility. Rather than having unequal numbers of quarks and antiquarks, partial separation occurred so that some parts of the Universe had a slight excess of quarks (including the part we live in) while other parts had a slight excess of antiquarks. That would raise the possibility that some parts of the Universe are made of antimatter. If that were so then we must hope that our galaxy stays well clear of those parts!

Time about 10^{-4} s

Quark–antiquark production had ceased and the quarks combined together in pairs to give different kinds of mesons or in threes to give baryons including neutrons and protons. Although protons and neutrons were available no atomic nuclei could be produced because it was far too hot and the large kinetic energies of the protons and neutrons overcame any binding energy that would enable them to combine.

It was mentioned in Section 1.4 that various kinds of exotic particles, that were unrelated to normal matter, have been detected in high-energy experiments. These would have existed in this period of the development of the Universe and the conditions then, after quark production, is what the particle physicists have managed to reproduce

Time about 1 s

A property of isolated neutrons is that they are unstable and, with a half-life of about 15 minutes, they decay into a proton, electron and antineutrino. On the other hand protons are stable and in isolation survive indefinitely. However, collision with another particle can disintegrate a proton; one way is by collision with an antineutrino that will give rise to a neutron and a positron. Another way is by collision with an electron, giving a neutron and neutrino. The ways in which neutrons and protons can break down are illustrated in Figure 1.18.

For proton disintegration the energies of the colliding particles have to be extremely high and as the Universe expanded and cooled down so the availability of high-energy impactors reduced. By contrast the disintegration of neutrons to give protons went on

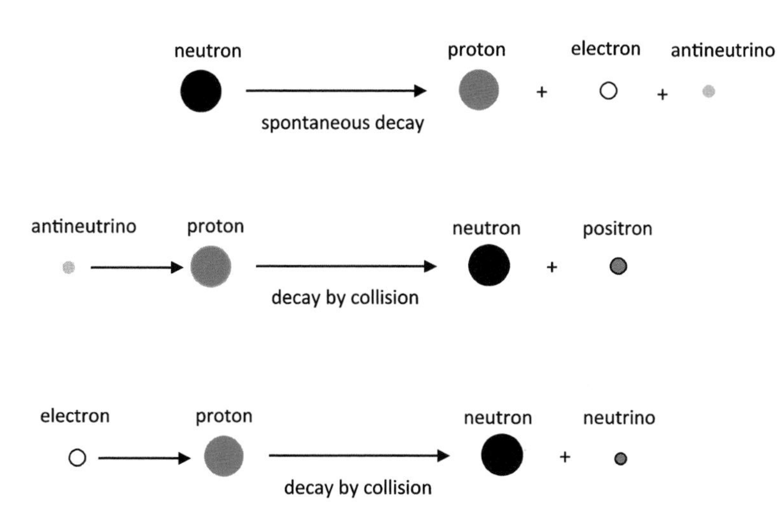

Figure 1.18. The spontaneous breakdown of an isolated neutron and the breakdown of a proton by two processes depending on collisions.

unabated — it was a spontaneous process. Consequently, during this period the ratio of protons to neutrons steadily increased.

Time about 100 s

Prior to this time, protons and neutrons had been moving so quickly that they did not stay close to each other long enough for bonding to take place. If they did form an association then it would only be temporary as a collision by a high-energy particle would cause it to break it up again. By now the temperature had fallen to a level where permanent associations of protons and neutrons could take place to form nuclei of some lighter elements — helium (He), a small amount of lithium (Li) and the hydrogen isotope deuterium (D). These nuclei are illustrated in Figure 1.19. By this time, due to the decay of isolated neutrons, there were about seven times more protons than neutrons and as the light elements formed that ratio increased as, proportionately, neutrons were reduced more rapidly than protons and more neutrons than protons were involved in producing lithium. Any neutrons left over after the production of the light nuclei would have decayed to increase the number of protons.

Time 10,000 years

By this time the Universe had cooled to such an extent that photons energetic enough to produce even the lightest particles were no longer available. The energy equivalent of the amount of matter that had been produced exceeded the energy associated with radiation. Matter and radiation were now independent of each other and as

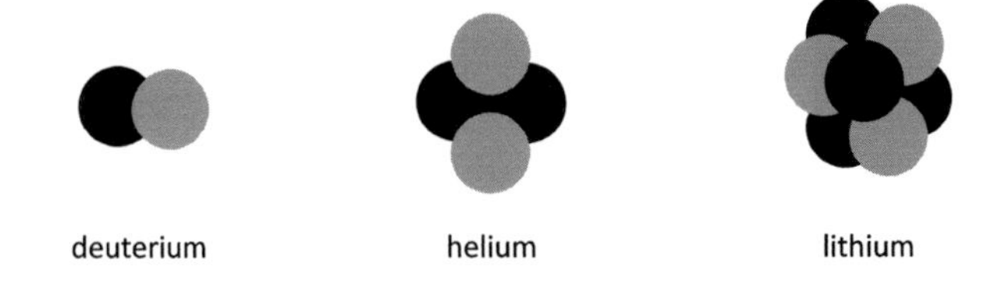

Figure 1.19. Light nuclei formed early in the expansion of the Universe (grey = proton, black = neutron).

the Universe expanded so the density of matter and the density of radiation both decreased. The amount of radiation in the Universe stayed constant thereafter but as its density decreased so did the temperature.

Time 500,000 years

The Universe abounded in light-atom nuclei including many protons, the nuclei of hydrogen atoms, but in prior times the temperature had been far too hot for electrons to attach themselves to nuclei — their kinetic energies were too high. At this time the temperature was low enough for electrons to bind to nuclei to form atoms. Those that attached themselves to protons produced hydrogen that was, and still is, the dominant element in the Universe.

From the Big Bang starting point, with no space, no time and no matter, but with an enormous amount of energy concentrated at a point, within 500,000 years the Universe evolved to a state where there was an abundance of matter and both space and time became tangible entities. The Universe was not one that we would recognize. It had a temperature of 60,000 K, compared with 2.73 K at present, and there was no material present of the type required to form terrestrial planets or the cores of major planets. The continuing expansion of the Universe gave the temperature we have today but other things have had to have happened to give us the present abundances of elements in the Earth and in the Universe. That is where stars have played a major role.

1.6. Dark Matter and Dark Energy

A feature of the Universe that we have not touched on is concerned with the mass that it contains. We interpret the motions of bodies in the Universe in terms of gravitational forces and Newtonian mechanics. Our galaxy, the Milky Way, is rotating and all the bodies within it orbit around its centre under the influence of the gravitational forces due to all other bodies — mainly those interior to itself. Estimating how much mass should be inside the Sun's orbit to explain its orbital period gives the surprising result for the Sun, and other stars in the galaxy, that the mass in the Milky Way, as judged by the light emitted by the stars it contains, accounts for less than 10% of the mass required to explain the rate of rotation. This result is confirmed for the motions of galaxies within clusters of galaxies. The observed mass is insufficient to hold the cluster together — with their observed relative velocities the galaxies should be flying apart. This has led to the idea of *missing mass*. Most of the mass in the Universe cannot be detected and much effort is going into finding out what it is. One theory is that they are particles, massive on the scale of fundamental particles, which interact so feebly with ordinary matter that it is virtually impossible to detect them. These particles have been called WIMPs (Weakly Interacting Massive Particles). Another theory, somewhat less favoured, is that it consists of ordinary matter but in an invisible form, either black holes or bodies that are cool and so emit little radiation, So far neither theory has been identified as the total contributor to the missing mass.

Another observation, based on studies of distant galaxies, is that the expansion of the Universe may be accelerating, with the recession

Figure 1.20. Saul Perlmutter.

speeds of galaxies from each other increasing with time (Saul Perlmutter, b. 1959, Figure 1.20, Nobel Prize in Physics, 2011). The reverse would be expected since masses should be pulling in rather than pushing out. This goes completely against what straightforward theory and instinct indicates. If galaxies are accelerating then the total energy of the observed Universe must be increasing so it is posited that there is some source for this extra energy — so-called *dark energy*; nobody has any idea what it could be.

As Albert Einstein showed, mass and energy are equivalent and the mass–energy content of the Universe, as presently estimated is:

Observed matter 4.9%

Dark matter 26.8%

Dark energy 68.3%

These strange dark quantities are of immense interest to cosmologists but do not impact on our immediate concern — the origin and properties of stars.

Problems 1

1.1 Three compounds of nitrogen and oxygen are formed with the mass of oxygen per 100 gm of nitrogen respectively 57 gm, 114 gm and 228 gm. What are possible structures for these compounds, given that it is known that an oxygen atom is more massive than a nitrogen atom? What is the estimated ratio of the mass of oxygen to nitrogen?

1.2 In a Thomson-type experiment to determine e/m for electrons the beam of electrons is undeflected with an electric field of 1,530 V m^{-1} and magnetic field of 1.332×10^{-4} T.

(i) What is the speed of the electrons?

(ii) With just the electric field operating the electron beam, moving midway between the plates, just misses the edge of the positive plate as it leaves the electric field region, which is 0.1 m in extent, with distance between the plates 0.02 m. What is the value of e/m for these electrons?

(iii) From special relativity theory, the mass of a body moving at speed v relative to an observer is

$$m_v = \frac{m_0}{\sqrt{1 - v^2/c^2}},$$

where m_0 is the rest-mass of the body and c the speed of light (2.998×10^8 m s^{-1}). From your result in part (ii) what is the value of e/m_0?

1.3 In a replica Millikan experiment the plates were 4 mm apart, the oil had a density of $1{,}109 \times 10^3$ kg m^{-3}, the acceleration due to gravity 9.80 m s^{-2} and the viscosity of air under the prevailing conditions was 1.511×10^{-5} kg m^{-1} s^{-1}. To suspend a drops the voltage across the plates was 232.4 V and when the electric field was removed the drop fell with a terminal speed of 1.142×10^{-4} m s^{-1}. What was the charge on the oil drop? How many electron charges are there on the oil drop?

1.4 The Sun emits 3.83×10^{26} W of radiation, all generated by nuclear reactions in its core. At what rate is it losing mass? What proportion of its mass will it lose in one billion years? (The Sun's mass is 1.989×10^{30} kg, the speed of light 2.998×10^8 m s^{-1} and there are 3.156×10^7 s in a year).

Part 2

Making Stars

Chapter 2

Some Useful Physical Theory

There are many theoretical results in physics and mechanics that apply over a wide range of astronomical scenarios, varying in scale from the behaviour of large clouds of gas in the Universe to the formation of planets. Here we deal with ten examples of such theory, which are applicable to aspects of stellar formation.

2.1. The Gravitational Potential Energy of a Spherically-Symmetric Sphere

Many astronomical bodies are of approximately spherical shape, any substantial departure usually being due to rotation that flattens the body along the rotation axis to the form of an oblate spheroid. However, the departure from spherical form is often small and here we shall assume that the bodies we are dealing with are perfect spheres. Another assumption is that the density variation within the sphere is dependent only on the distance from the centre, which makes the body spherically symmetric. In the analysis that follows we shall use the result that the gravitational influence of a spherically-symmetric sphere on an exterior body is the same as that of the total mass of the sphere situated at its centre.

The gravitational potential energy of a body is the work done *against* gravity in assembling the body from material that was originally an infinite distance away. Since in that scenario gravity is doing the work of assembly, the work done against the field is negative, as is the gravitational potential energy of the body. Another way of defining it is to say that it is the negative of the energy required to disperse the material of the body to an infinite distance.

Consider a partially-assembled spherically-symmetric body of current radius r has mass given by

$$M(r) = 4\pi \int_0^r \rho(x)x^2 dx. \tag{2.1}$$

If new shells of thickness dr are successively added then the work done against the field in producing a sphere of radius R is

$$\Omega = -4\pi G \int_{r=0}^{R} \frac{M(r)\rho(r)r^2}{r} dr = -4\pi G \int_{r=0}^{R} M(r)\rho(r)r dr, \tag{2.2}$$

where G is the *gravitational constant.*[1]

For a uniform sphere of density ρ

$$M = \frac{4}{3}\pi\rho r^3, \tag{2.3}$$

so that

$$\Omega = -\frac{16\pi^2}{3}G\rho^2 \int_{r=0}^{R} r^4 dr = -\frac{16\pi^2}{15}G\rho^2 R^5 = -\frac{3}{5}\frac{GM^2}{R}. \tag{2.4}$$

For any spherically-symmetric distribution of density the potential energy is of the form $\alpha GM^2/R$, in which α is a numerical constant, the value of which depends on the density distribution.

As an example we calculate the gravitational potential energy of a sphere of radius R within which $\rho(x) = a/x$. This gives an infinite density at the origin but we may take the density equation as an approximation where the density just goes to a large value at the origin. From (2.1) we calculate the mass out to radius r. This is

$$M(r) = 4\pi \int_0^r \frac{a}{x} x^2 dx = 2\pi a r^2. \tag{2.5}$$

[1]The gravitational force between point masses m_1 and m_2 distance d apart is Gm_1m_2/d^2 where $G = 6.674 \times 10^{-11}\ \mathrm{m^3\,kg^{-1}\,s^{-2}}$.

Inserting this into (2.2)

$$\Omega = -8\pi^2 a^2 G \int_0^R r^2 dr = -\frac{8}{3}\pi^2 a^2 G R^3.$$

Expressing in terms of

$$M(R) = 2\pi a R^2$$

$$\Omega = -\frac{2}{3} G \frac{M(R)^2}{R}. \tag{2.6}$$

2.2. The Virial Theorem

The Virial Theorem applies to a system of particles with pair interactions for which, although the particles are moving, the distribution of particles, in a statistical sense, does not vary with time. The theorem, introduced by Rudolf Clausius (1822–1888), a German physicist and mathematician (Clausius, 1870) states that

$$2K + \Omega = 0, \tag{2.7}$$

where K is the total translational kinetic energy of the particles and Ω is the potential energy of the system.

A very simple illustrative example of the validity of the theorem is that of a body of mass m in a circular orbit around a body of mass M, where $M \gg m$. The kinetic energy of the system is $K = \frac{1}{2}mv^2$, where v is the orbital speed. Equating centrifugal force to the gravitational force on m,

$$\frac{mv^2}{r} = \frac{GMm}{r^2},$$

where r is the radius of the orbit. Substituting for v^2 in the expression for K gives $K = \frac{GMm}{2r}$. Since the potential energy $\Omega = -\frac{GMm}{r}$ this gives $2K + \Omega = 0$,

Here we shall show the general validity of the theorem for a system of gravitationally interacting bodies. We consider a system of N bodies for which the ith has mass m_i and coordinates (x_i, y_i, z_i).

We define the *geometrical moment of inertia* as

$$I = \sum_{i=1}^{N} m_i \left(x_i^2 + y_i^2 + z_i^2\right). \tag{2.8}$$

Differentiating I twice with respect to time and dividing by two we find

$$\frac{1}{2}\ddot{I} = \sum_{i=1}^{N} m_i \left(\dot{x}_i^2 + \dot{y}_i^2 + \dot{z}_i^2\right) + \sum_{i=1}^{N} m_i (x_i \ddot{x}_i + y_i \ddot{y}_i + z_i \ddot{z}_i). \tag{2.9}$$

The first term on the right-hand side, the sum of the mass times speed squared for each particle, is $2K$. The second term can be transformed by noting that $m\ddot{x}$ is the x component of the total force on the body i due to all the other particles, so that

$$m_i x_i \ddot{x}_i = \sum_{\substack{j=1 \\ \mathrm{i} \neq i}}^{N} G m_i m_j \frac{x_i (x_j - x_i)}{r_{ij}^3}, \tag{2.10}$$

where $r_{ij} = \{(x_i - x_j)^2 + (y_i - y_j)^2 + (z_i - z_j)^2\}^{1/2}$ is the distance between particles i and j. The factor $(x_j - x_i)/r_{ij}$ in (2.10) gives the x-component of the force due to particle j on particle i.

In producing the second summation on the right-hand side of (2.9), for every term as shown in the summation in (2.10) involving (i, j) there is another term involving (j, i), which is

$$G m_i m_j \frac{x_j (x_i - x_j)}{r_{ij}^3}.$$

Combining these two terms, the second term on the right-hand side of (2.9) becomes

$$\begin{aligned}
&\sum_{i=1}^{N} m_i (x_i \ddot{x}_i + y_i y_i + z_i \ddot{z}_i) \\
&= - \sum_{pairs} G m_i m_j \frac{(x_i - x_j)^2 + (y_i - y_j)^2 + (z_i - z_j)i^2}{r_{ij}^3} \\
&= - \sum_{pairs} \frac{G m_i m_j}{r_{ij}} = \Omega,
\end{aligned} \tag{2.11}$$

which is the gravitational potential energy. Equation (2.9) now appears as

$$\frac{1}{2}\ddot{I} = 2K + \Omega. \tag{2.12}$$

Although particles move, if the system stays within the same volume with the same general distribution of matter, at least in a time-averaged sense, then the expected value of $\ddot{I} = 0$ and the virial theorem is verified. The Virial Theorem has a wide range of applicability and can be applied to the motions of stars within a cluster of stars or to an individual star where the translational kinetic energy is that due to the thermal motion of the particles that form the star.

As an example of the first application we find the mean-square speed for a system of 100 stars, of average mass 1.6×10^{30} kg maintaining a uniform distribution within a sphere of radius 10^{12} m. The total mass of stars is 1.6×10^{32} kg and if we assume that they approximate to a uniform distribution within the sphere then the gravitational potential energy is

$$\Omega = -0.6 \times G\frac{\mathrm{M}^2}{\mathrm{R}} = -0.6 \times 6.674 \times 10^{-11}$$
$$\times \frac{(1.6 \times 10^{32})^2}{10^{12}} = -1.025 \times 10^{42}.$$

The translational kinetic energy is
$K = 1.6 \times 10^{32}\, \overline{v^2}$ where $\overline{v^2}$ is the mean-square speed.

From the Virial Theorem
$2 \times 1.6 \times 10^{32}\, \overline{v^2} = 1.025 \times 10^{42}$ giving

$\bar{v^2} = 3.204 \times 10^6\,\mathrm{m^2\,s^{-2}}$ that corresponds to a root-mean-square speed of about $1.8\,\mathrm{km\,s^{-1}}$.

2.3. The Jeans Critical Mass

Many astronomical bodies begin their identifiable existence as large approximately-spherical gaseous objects. If a gaseous sphere is formed then there are two influences acting on it in opposite directions. The first of these is the force of gravity that, pulling inwards on the material, which either tends to hold the sphere together or causes it to collapse. The second is the kinetic energy associated with the motion of the particles that constitute the material of the sphere, which tends to cause it to fly apart. If these two influences are in balance then the sphere is in an equilibrium state. For a static sphere of gaseous material at a particular density and temperature, if the mass is below a certain limit then the kinetic energy will dominate and the sphere will disperse while, for a mass above the limit, gravity will dominate and the sphere will collapse. This limiting mass, known as the *Jeans critical mass* (Jeans, 1902), was first described by the British astrophysicist, James Jeans (1877–1948; Figure 2.1).

One approach to find the Jeans critical mass is to consider that it corresponds to the situation when the Virial Theorem (Section 2.2) is just satisfied. If the right-hand side of (2.12) is positive then the geometrical moment of inertia, and hence the mean value of r_{ij}^2, is increasing. This indicates that the material is expanding and hence moving outwards and dispersing. Conversely, if the right-hand side of (2.12) is negative then the sphere will be in a state of collapse.

Figure 2.1. James Jeans.

Making the left-hand side equal to zero, the usual form of the Virial Theorem, gives the critical equilibrium state.

For a static sphere, where there is no mass motion of the material, then the only source of kinetic energy is the thermal motion of the molecules. The part of the kinetic energy of the molecules that is relevant here is the *translational* component. According to statistical mechanics, the mean kinetic energy is $1/2\,kT$ for each degree of freedom; for translational kinetic energy there are three degrees of freedom, corresponding to motion in the x-, y- and z-directions; k is Boltzmann's constant[2] and T the absolute temperature. Molecules, with more than a single atom in each entity, also have kinetic energy associated with tumbling modes of motion, giving other degrees of freedom, but this energy is not relevant in the present context. If the total mass of the gas sphere is the Jeans critical mass, M_{Jc}, then its total translational kinetic energy is

$$K = \frac{3kTM_{Jc}}{2\mu}, \tag{2.13}$$

where μ is the mean mass of the molecules forming the gas.

The gravitational potential energy of the sphere will depend on the distribution of matter within it, which we will take to be spherically symmetric. This has the general form

$$\Omega = -\alpha\frac{GM_{Jc}^2}{R}, \tag{2.14}$$

where α is a numerical constant, equal to 0.6 for a uniform sphere, and R is the radius. Using the Virial Theorem for a uniform sphere, the Jeans critical mass is found from

$$2K + \Omega = \frac{3kTM_{J_c}}{\mu} - 0.6\frac{GM_{J_c}^2}{R} = 0,$$

giving

$$M_{Jc} = \frac{5k}{\mu G}RT. \tag{2.15}$$

[2]The Boltzmann constant is $1.381 \times 10^{-23}\,\mathrm{J\,s^{-1}}$.

Substituting $R = \left(\frac{3M_{Jc}}{4\pi\rho}\right)^{1/3}$ and rearranging gives

$$M_{Jc} = \left(\frac{375k^3}{4\pi\mu^3G^3}\right)^{1/2}\left(\frac{T^3}{\rho}\right)^{1/2} = Z\left(\frac{T^3}{\rho}\right)^{1/2}. \tag{2.16}$$

For a mix of gases consisting of atomic hydrogen, molecular hydrogen and helium, a value of μ around 4×10^{-27} kg gives $Z = 2.0 \times 10^{21}$ in SI units. The corresponding Jeans critical masses for various combinations of temperature and density are shown in Figure 2.2.

2.4. Free-Fall Collapse

The model we are taking here to illustrate free-fall collapse is that of a uniform sphere of mass M, initial radius, r_0, and initial uniform density, ρ, collapsing from rest under the force of gravity alone. If the

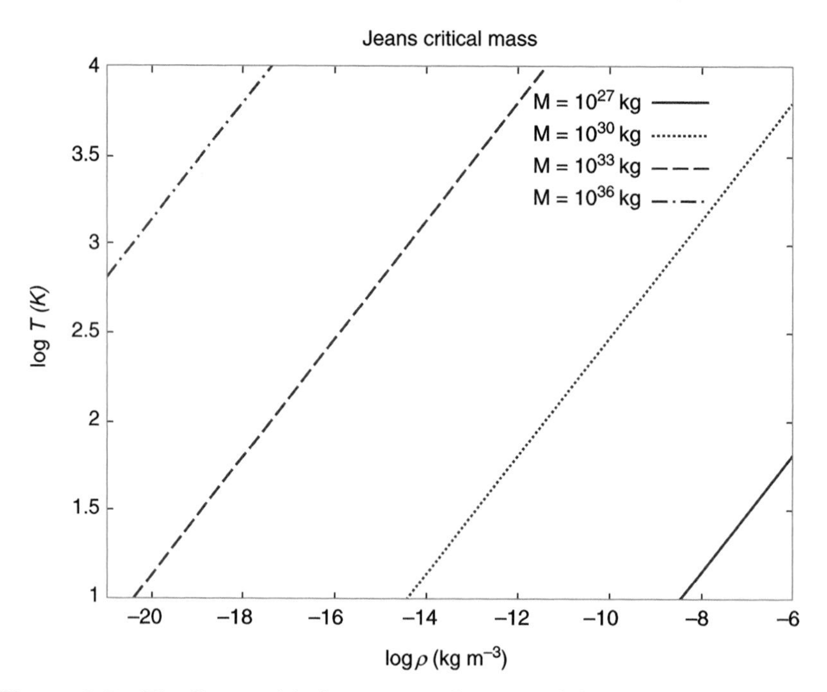

Figure 2.2. The Jeans critical mass as a function of density and temperature for $\mu = 4.0 \times 10^{-27}\,\mathrm{kg\,m^{-3}}$.

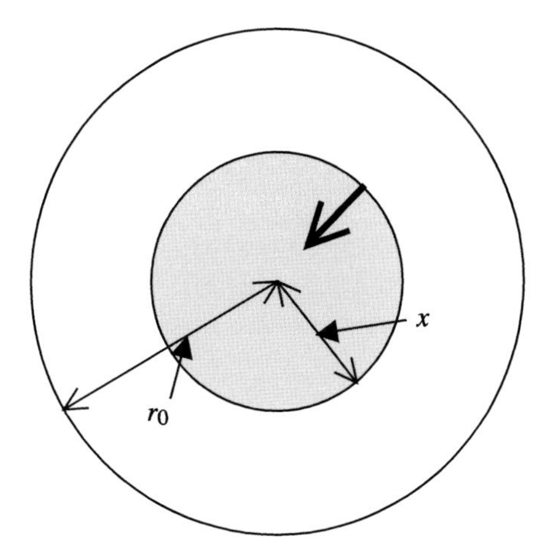

Figure 2.3. The shaded volume provides the gravitational force acting at distance from the centre.

material of such a sphere were a gas then clearly there would be pressure forces acting as well, so the model is better thought of as small gravitationally-interacting solid bodies uniformly distributed throughout the sphere.

In Figure 2.3 we show the sphere at the beginning of the collapse. A particle at a distance x from the centre experiences a force due to the shaded volume that gives it an acceleration

$$\frac{d^2x}{dt^2} = -\frac{\frac{4}{3}\pi x^3 \rho G}{x^2} = -\frac{4}{3}\pi\rho G x. \tag{2.17}$$

This equation is linear in x, which implies that the acceleration and also the velocity at different points at any time will be proportional to the distance of the point from the centre. Hence during the collapse the density increases but remains uniform throughout the sphere.

Since there are no dissipative forces acting, the total energy of any particle must remain constant throughout the collapse. Equating the potential energy at the beginning of the collapse to the total energy (potential plus kinetic) at some subsequent time, for a boundary particle we have

$$-\frac{GM}{r_0} = -\frac{GM}{r} + \frac{1}{2}\left(\frac{dr}{dt}\right)^2, \tag{2.18}$$

where r is the radius of the sphere at time t. Rearranging we have

$$\frac{dr}{dt} = -\left\{2GM\left(\frac{1}{r} - \frac{1}{r_0}\right)\right\}^{\frac{1}{2}}, \tag{2.19}$$

where the negative sign is taken because the sphere is collapsing.

Substituting $r = r_0 \sin^2\theta$ gives

$$\frac{d\theta}{dt} = -\left(\frac{GM}{2r_0^3}\right)^{\frac{1}{2}} \frac{1}{\sin^2\theta}. \tag{2.20}$$

If at time t the radius of the sphere is r_t then, from (2.20)

$$t = -\left(\frac{2r_0^3}{GM}\right)^{\frac{1}{2}} \int_{\pi/2}^{\theta_t} \sin^2\theta d\theta, \tag{2.21}$$

where $\theta_t = \sin^{-1}\left\{\left(\frac{r_t}{r_0}\right)^{1/2}\right\}$.

Evaluating the definite integral in (2.21) and then substituting back in terms of r gives

$$t = \left(\frac{r_0^3}{2GM}\right)^{\frac{1}{2}} \left\{\frac{\pi}{2} - \sin^{-1}\left[\left(\frac{r_t}{r_0}\right)^{\frac{1}{2}}\right] + \left(\frac{r_t}{r_0}\right)^{\frac{1}{2}}\left(1 - \frac{r_t}{r_0}\right)^{\frac{1}{2}}\right\}. \tag{2.22}$$

The time for complete collapse — the free-fall time t_{ff} — is found by making $r_t = 0$ in (2.22) giving

$$t_{ff} = \frac{\pi}{2}\left(\frac{r_0^3}{2GM}\right)^{\frac{1}{2}} = \left(\frac{3\pi}{32\rho G}\right)^{\frac{1}{2}}. \tag{2.23}$$

From (2.22) and (2.23)

$$\frac{t}{t_{ff}} = 1 - \frac{2}{\pi}\left\{\sin^{-1}\left[\left(\frac{r_t}{r_0}\right)^{\frac{1}{2}}\right] - \left(\frac{r_t}{r_0}\right)^{\frac{1}{2}}\left(1 - \frac{r_t}{r_0}\right)^{\frac{1}{2}}\right\}. \tag{2.24}$$

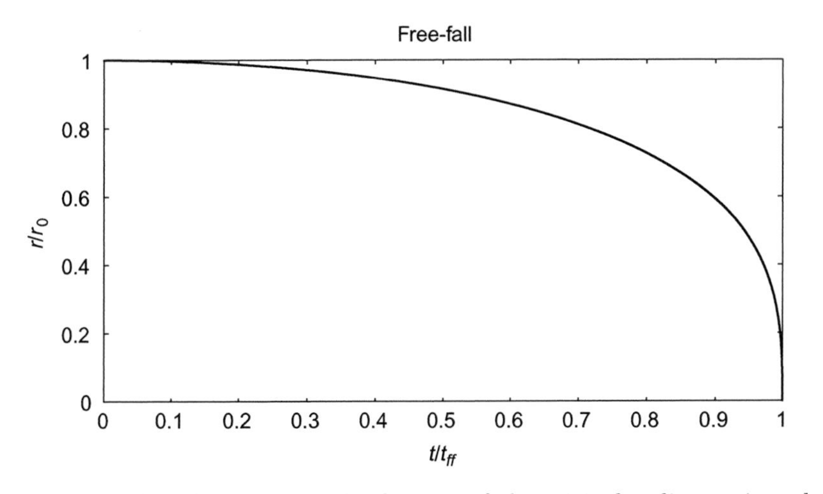

Figure 2.4. Free fall, showing the fraction of the original radius against the fraction of the free-fall time.

Figure 2.4 shows the variation of r_t/r_0 with time expressed in units of the free-fall time. Even for a collapse of solid particles, so that pressure forces do not occur, the final radius cannot be zero since that would imply infinite density. However, as is seen from the figure, the collapse is slow at first, the radius decreasing by less than 9% after 50% of the free-fall time. Thereafter the collapse speeds up and then is very rapid at the end. The difference of time between collapsing to zero radius and to a small finite radius is negligible.

For an initially uniform gas sphere the pressure forces will be small at first because there will be no pressure gradient except at the boundary. Thus the first part of the collapse of a gas sphere will approximate to free fall. If the heat energy generated by the compression of the gas is efficiently radiated away then the free-fall time may not be a serious underestimate of the time to collapse to high density.

2.4.1. *The Relationship of the Virial Theorem to Free Fall*

The rates of collapse found for the protostar in Problems 2.3(b) and 2.4(a) are very different and the difference is due to the neglect of pressure forces in the latter problem. Pressure has dimensions 'energy per unit volume' so that, for example, the pressure of a perfect gas is given by

$$p = NkT, \tag{2.25}$$

where N is the number of molecules per unit volume and kT is two-thirds of their average translational kinetic energy. When the Virial Theorem is applied to a cluster of moving stars then the motion of the stars gives an effective translational kinetic energy per unit volume that acts as a pressure. On the other hand, if the collapse of a uniform spherical distribution of static massive objects were considered then it would be treated as a free-fall problem, which is equivalent to using equation (2.7) with $K = 0$. To show this we note that the geometrical moment of inertia for a uniform sphere of radius R and mass M is given by

$$I = \frac{3}{5}MR^2,$$

giving

$$\frac{1}{2}\ddot{I} = \frac{3}{5}M\dot{R}^2 + \frac{3}{5}MR\ddot{R}. \tag{2.26}$$

The gravitational potential energy of the sphere is

$$\Omega = -\frac{3}{5}\frac{GM^2}{R}.$$

Initially $\dot{R} = 0$ so using (2.26) and (2.12)

$$\frac{1}{2}\ddot{I} = \frac{3}{5}MR\ddot{R},$$

giving, without the kinetic energy term

$$\frac{3}{5}MR\ddot{R} = -\frac{3}{5}\frac{GM^2}{R},$$

or

$$\ddot{R} = -\frac{GM}{R^2},$$

the expected result in the absence of pressure forces.

In scientific work the free-fall time for complete collapse is sometimes given as an approximation to the time it would take a protostar (see Problem 2.3) to collapse to a condensed form. This would only be valid if the kinetic energy is much less than the magnitude of the potential energy, corresponding to a low temperature and a sufficiently massive protostar.

2.5. Gravitational Instability

The application, and theoretical treatment, of gravitational instability was given by James Jeans in 1917 in relation to a theory of formation of the Solar System. He envisaged that a massive star had passed close to the Sun and drawn from it a filament of solar material. He then showed that the filament would be gravitationally unstable and break up into a string of condensations. The solar filament would have been mainly gas but with some solid component. For that scenario the density and temperature would almost certainly have been different in different parts of the filament. However, for our present consideration it will be assumed that the filament has a uniform density and temperature

A simple argument, illustrated in Figure 2.5, illustrates why the filament will break up. Figure 2.5(a) shows a perturbation of the filament in the form of a small density excess in region A. Because of an imbalance of forces, material near A experiences an attraction towards A and this creates two lower density regions, B and B′, on either side of A (Figure 2.5(b)). Material beyond B and B′, at C and C′, now experiences outward accelerations and produces high density regions at D and D′ (Figure 2.5(c)). These high-density regions act like the original perturbation at A and so the wave-like disturbance of the filament travels outwards.

A formal analysis of this model in terms of the properties of the gas gives the distance, l, between the high-density condensations in the stream, but the general form of the expression can be found just from dimensional analysis. The rate at which a disturbance moves along the filament is related to the speed of sound in the gas given by

$$c = \sqrt{\frac{\gamma k T}{\mu}}, \tag{2.27}$$

in which γ is the usual ratio of specific heats of the gas, k is the Boltzmann constant and T and μ are respectively the temperature and mean molecular weight of the gas. Other factors influencing l are the gravitational constant, G, and the density of the gas, ρ. The relationship found by Jeans, with the numerical constant not given by dimensional analysis, is

$$l = \left(\frac{\pi}{\gamma G \rho}\right)^{\frac{1}{2}} c = \left(\frac{\pi k T}{G \rho \mu}\right)^{\frac{1}{2}}. \tag{2.28}$$

In the Jeans analysis the stream is found to have a periodic variation of density and l is the wavelength of the density fluctuation. Something the analysis does not include is the line-density of the filament, σ, i.e. the mass per unit length, a quantity that does not control the wavelength of the density wave. If σ is low then the blobs formed in the filament will have less than the Jeans critical mass and disperse; if σ is sufficiently large then the blobs will have more than the Jeans critical mass and collapse.

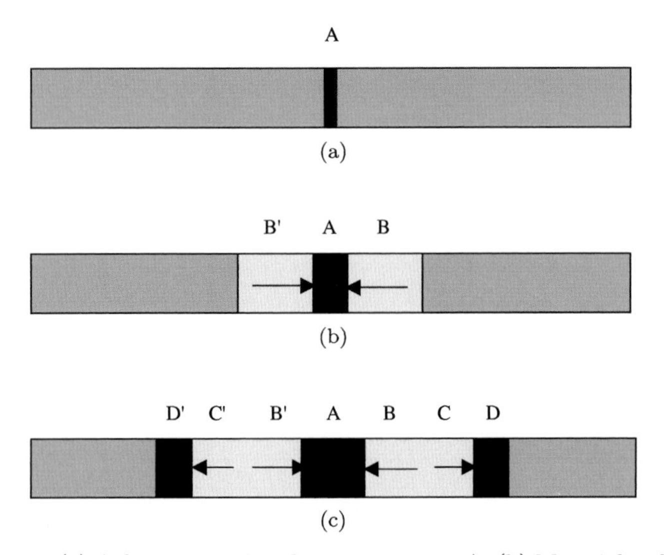

Figure 2.5. (a) A filament with a density excess at A. (b) Material at B and B′ is attracted towards A. (c) Material at C and C′ moving away from the depleted regions at B and B′ create higher density regions at D and D′.

Gravitational instability can also occur in two and three dimensions. Assuming that l in equation (2.28) is the diameter of a spherical condensation that will result from the gravitational instability of a gas cloud then the mass of the condensation is

$$M_{\text{Jc}} = \frac{4}{3}\pi\rho\left(\frac{l}{2}\right)^3 = \frac{\pi^{5/2}}{6}\left(\frac{k^3T^3}{\mu^2G^3\rho}\right)^{1/2} = 2.92\left(\frac{k^3T^3}{\mu^3G^3\rho}\right)^{1/2}, \tag{2.29}$$

which is similar to the Jeans critical mass that is, from equation (2.16),

$$M_{\text{Jc}} = \left(\frac{375}{4\pi}\right)^{1/2}\left(\frac{k^3T^3}{\mu^3G^3\rho}\right)^{1/2} = 5.46\left(\frac{k^3T^3}{\mu^3G^3\rho}\right)^{1/2}. \tag{2.30}$$

The coefficients of the expressions in (2.29) and (2.30) depend somewhat on the theoretical approach being used to derive them so to take the condensations that are produced by three-dimensional gravitational instability as having the Jeans critical mass is a reasonable assumption.

2.6. The Equipartition Theorem

A very well-known distribution, which comes from the field of statistical mechanics, is the Maxwell–Boltzmann distribution. If, for example, there is an ideal gas in a container at temperature T then the individual molecules, each of mass m, will move freely, except for occasional elastic collisions, (i.e. collisions with no loss of mechanical energy) and their speeds will follow the distribution

$$f(v) = 4\pi v^2\left(\frac{m}{2\pi kT}\right)^{\frac{3}{2}}\exp\left(-\frac{mv^2}{2kT}\right). \tag{2.31}$$

The function is normalized so integrating between $v = 0$ and $v = \infty$ gives unity. The meaning of the distribution is that the proportion of molecules with speeds between v and $v + dv$ is $f(v)dv$. This can be converted into a distribution of kinetic energy, K, by the substitution

$$v = \sqrt{\frac{2K}{m}} \quad \text{and} \quad dv = \sqrt{\frac{1}{2Km}}\, dK.$$

Giving

$$f(K)dK = 2\left(\frac{1}{kT}\right)^{3/2}\left(\frac{K}{\pi}\right)^{1/2}\exp\left(-\frac{K}{kT}\right)dK. \tag{2.32}$$

Equation (2.32) shows that the distribution of kinetic energy is independent of the mass of the particle and only dependent on the temperature. This conclusion is also true if there is a mixture of molecules of different masses; each type of molecule will take up the distribution (2.32). Thus if there is a mixture of molecules of masses m_1 and m_2 then their mean kinetic energies are equal. This is the *equipartition theorem*. Since the mean kinetic energies are equal

$$\overline{K} = \frac{1}{2}m_1\overline{v_1^2} = \frac{1}{2}m_2\overline{v_2^2}. \tag{2.33}$$

What is true for the mean square speeds is also true for the mean speeds squared so we find

$$\frac{\overline{v_1}}{\overline{v_2}} = \sqrt{\frac{m_2}{m_1}}. \tag{2.34}$$

The equipartition theorem applies to particles other than molecules. For example, in an environment where there is plasma — a mixture of electrons and atomic ions — the electrons will be moving with a much greater mean speed. This difference of speed is important in some astronomical contexts.

2.7. Cooling Processes

Material in the galaxy is being constantly heated by the radiation coming from stars and also by *cosmic rays*, which are mostly high-energy particles — nearly all hydrogen but with atoms of almost all other elements also present. Unless there were some compensating cooling mechanisms all material would have every-increasing temperature, something that does not happen because cooling mechanisms are also operating. One process was described by the eminent Japanese astrophysicist, Chushiro Hayashi (1920–2010), who showed in 1966 that dust particles act like radiators, such as those of a car or central-heating system, and so dust cools the gas. However, as shown by Michael Seaton (1923–2007) in 1955, a much more effective cooling mechanism is due to the presence of molecules, atoms and ions, together with large numbers of free electrons, in the galactic material. Because of the equipartition theorem free electrons move very quickly and hence make many collisions with molecules, atoms and ions. The cooling mechanism, produced by the collision of a free electron with singly-ionized carbon atom is illustrated in Figure 2.6.

In Figure 2.6(a) a free electron approaches a carbon ion. In Figure 2.6(b) the free electron pushes an ion-electron into a higher energy state and consequently loses some of its own energy. Figure 2.6(c) shows the struck ion-electron spontaneously returns to its original state with the emission of a photon that leaves the local system with the speed of light. The net result is that the free electron has lost kinetic energy so that the average kinetic energy per particle in the system, and hence its temperature, is reduced. Cooling by singly-ionized carbon, written as C^+ to indicate its single positive charge, is very effective; a similar process can work with other ions, atoms and molecules. The rate-of-cooling equation for C^+ in the interstellar medium it is of the form

$$\frac{dQ}{dt} = 1.04 \times 10^{14} \rho T^{-1/2} \exp\left(-\frac{92}{T}\right) \mathrm{W\,kg^{-1}}, \qquad (2.35)$$

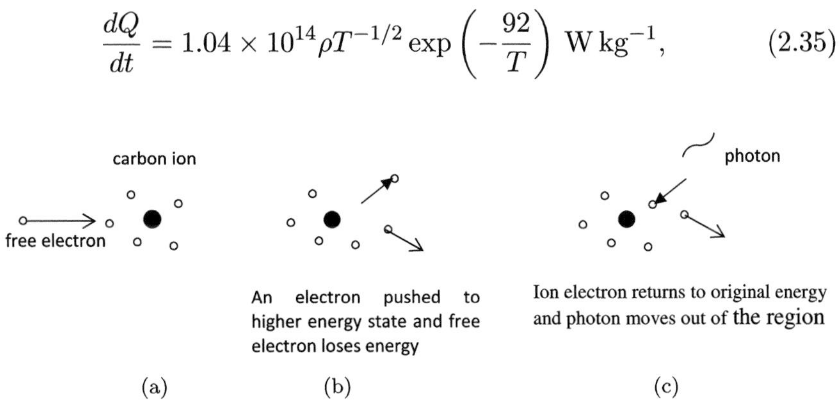

Figure 2.6. The cooling of a gaseous medium by free-electron collision with a carbon ion.

in which dQ/dt is the loss of thermal energy per unit mass per unit time, ρ is the density of the medium and T is the absolute temperature. The coefficient before ρ in (2.35) depends on the assumed composition of the ISM. Other effective coolants are ionic silicon and iron and there are several important atomic coolants, especially oxygen that is very effective, and molecular coolants, such as molecular hydrogen, H_2. Cooling processes play a critical role in stellar formation and evolution.

2.8. Opacity

A feature of the cooling processes described in the previous section is that it is assumed that photons, coming either from radiating dust or from the electron-excitation process, should be able to leave the region of the body of interest. Whether or not this can happen depends on a property of the material called *opacity*. It describes the resistance of the material to the flow of radiation through it. If a beam of radiation of intensity I travels a distance dx through an absorbing medium then the intensity is changed by an amount

$$dI = -\zeta I dx, \tag{2.36}$$

where ζ is the *linear absorption coefficient*, with dimensions m^{-1}. *Opacity*, a quantity preferred by astronomers, is defined as

$$\kappa = \zeta/\rho,$$

so that

$$dI = -\kappa\rho I dx. \tag{2.37}$$

Opacity has the dimensions $m^2\ kg^{-1}$ and expresses the resistance to the passage of radiation as an effectively opaque, i.e. impenetrable, area per unit mass of material in the path. The action of opacity is to subtract energy from the radiation and to convert it to some other form of energy. This can be by heating material, by exciting material in some way or by scattering the oncoming radiation.

For any mix of gasses the opacity is strongly dependent on temperature and for the usual mixture of hydrogen and helium that

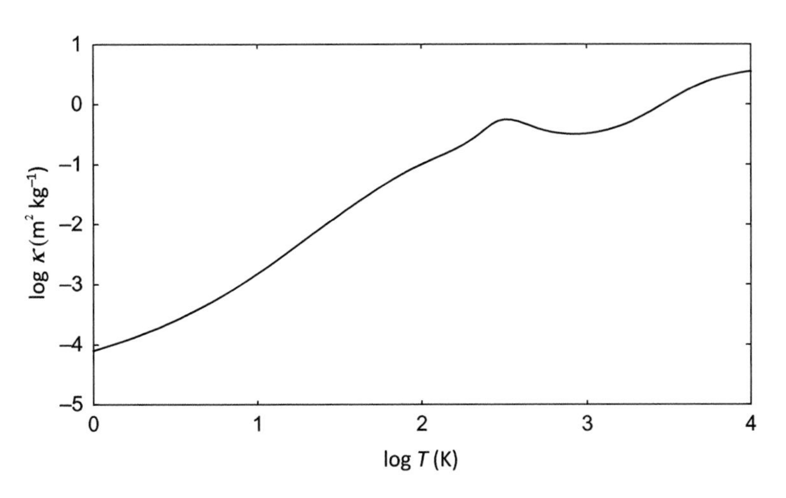

Figure 2.7. The Rosseland mean opacity.

form the major component of stars the *Rosseland mean opacity* is often used (Figure 2.7). Whether or not radiation can leave the body of interest also depends on its dimensions. For a given opacity and density, radiation might readily leave a very small body but be trapped within a large one. This brings into play another quantity known as the *optical depth* defined as

$$\tau = \ln\left(\frac{I_0}{I}\right), \tag{2.38}$$

where I_0 is the intensity of a beam of radiation falling on a body and I is the intensity of the transmitted beam. From (2.36) if $I = I_0$ when $x = 0$ and the thickness of the body is d we find $I_d = I_0 \exp(-\zeta d)$ from which

$$\tau = \zeta d = \kappa\rho d. \tag{2.39}$$

For any particular system, radiation will be generated over a volume and its pathway to escape will depend on where it is situated and the arrangement of boundaries. However, we may take it as a general rule that if D is a typical dimension of an object and $\kappa\rho D \gg 1$ the radiation will be trapped within it. Conversely if $\kappa\rho D \ll 1$ then the radiation will be readily lost. As an example we take a sphere of stellar-type gas of radius 10^{14} m, density 10^{-13} kg m^{-3} and temperature 1,000 K. The total mass of the sphere is 4.2×10^{29} kg, just over $0.2\,M_{\odot}$.[3] At a temperature of 1,000 K, $\kappa = 0.45\,\mathrm{m^2\,kg^{-1}}$ so that $\tau = 0.45 \times 10^{-13} \times 10^{14} = 4.5$. This is substantially greater than unity and little radiation will escape from the interior of the body.

2.9. The Light from Stars

Light is a form of electromagnetic radiation, which involves coordinated fluctuations in electric and magnetic fields, and these waves can have a wide range of wavelengths, illustrated in Figure 2.8. At the radio end the wavelength is 100 km while at the γ-ray end it is 10^{-13} m, one ten-thousand-millionth of a millimetre. At the blue end of the visible spectrum the wavelength is 400 nm[4] and at the red end 700 nm; it is interesting to note how little of the electromagnetic spectrum we can detect with our eyes!

In describing the temperature of bodies we sometimes use descriptions such as 'red-hot' or 'white hot', indicating that the colour of the light emitted by an object varies with its temperature. If we progressively heat a piece of iron at first it emits no light but if we hold a hand near the iron then we feel the heat coming from it. This heat radiation corresponds to the infrared region of Figure 2.8. Further heating the iron it glows a dull red colour; it is still emitting

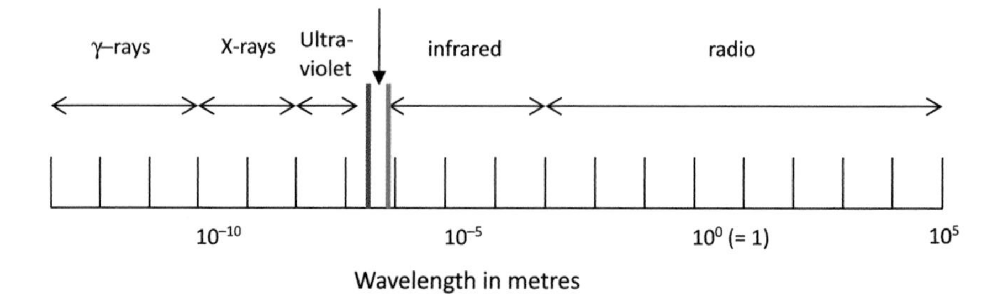

Figure 2.8. The electromagnetic spectrum.

[3] $M_{\odot}$ represents the mass of the Sun, 1.989×10^{30} kg.
[4] 1 nanometre (nm) is 10^{-9} m.

large quantities of heat radiation but it is also beginning to radiate in the red part of the visible spectrum. Heating further still the iron first becomes a brighter red, then yellow and then a brilliant white. The white colour indicates that there is substantial radiation coming from all parts of the visible spectrum; we can still detect radiated heat — a great deal of it — so there is also radiation coming from the infrared region. Something we will not be able to detect visually is the emission of ultraviolet radiation, coming from the region beyond the blue end of the spectrum. At an even higher temperature, the white light will be tinged with blue, indicating that the blue end of the spectrum is radiating more strongly than other parts of the visible region. What we learn from this simple experiment is that the higher the temperature the shorter is the average wavelength of the emitted radiation.

Before going further we must say something about the way that scientists define temperature. We are familiar with the Centigrade (Celsius) scale because it is used in weather reports. It is defined by taking the freezing point of water as 0°C and the boiling point of water as 100°C, both assuming normal pressure at ground level. A rather more fundamental way of defining temperature is by the average energy of the particles at a particular temperature. In a fluid — a gas or liquid — the energy manifests itself by the fluid particles moving within the space it occupies. For a solid, particles are bound to each other in a fixed configuration and in this case the energy manifests itself as vibrations of atoms around a fixed point. With this description of temperature it is possible to define an absolute zero of temperature, when the particles are not moving.[5] Then moving upwards from zero each one degree corresponds to the same increase of average energy per particle as is given by the Centigrade scale. This defines the Absolute, or Kelvin, scale of temperature. The unit of temperature is the *Kelvin* (symbol K) and temperature in kelvin is the temperature in centigrade +273.15.

[5]Because of the constraints of quantum mechanics, systems cannot have zero energy. The least energy they can have, the *zero-point energy*, is tiny and can be neglected in our description of temperature.

Hence on the absolute scale the freezing point of water is 273.15 K and the boiling point 373.15 K.

Figure 2.9 shows the form of the emission of radiation from bodies at three different temperatures, which correspond to possible temperatures of stars. These are Planck radiation curves, the derivation for the form of which is given in Appendix A. For 5,000 K the visible radiation is biased towards the red end of the spectrum and the net colour perception will be orangey-yellow. At 6,500 K the bias is towards the blue end and the result will be a slightly-bluish white, while at 8,000 K the white will be strongly tinged with blue. It seems therefore that the temperature of a star could be judged by its colour and, indeed, the three temperatures displayed in the figure are easily distinguishable. However, it would be very difficult to distinguish stars with temperatures differing by only 50 K in this way and astronomers have found a much better way if finding the temperature of stars, which will be described in Section 9.1.

Final points to note from Figure 2.9 is that, from a given area of a body, the amount of energy emitted, proportional to the area under the curve, increases rapidly with temperature and that the peak of the emission moves to shorter wavelengths as the temperature

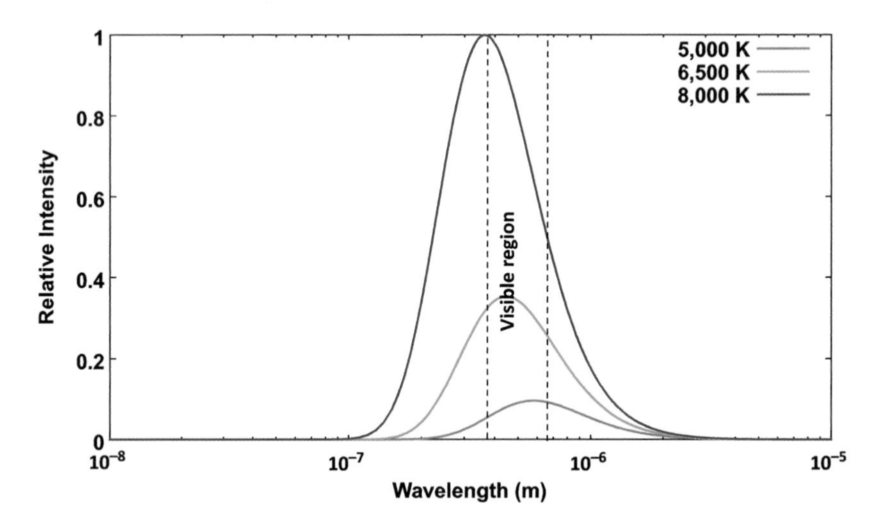

Figure 2.9. Relative radiation curves at 5,000 K, 6,500 K and 8,000 K.

increases. Theory shows that the energy radiated per unit area increases as T^4, where T is the absolute temperature, and that the peak wavelength, λ_{peak} is lower for higher temperatures, the relationship between λ_{peak} and T, known as *Wien's Law*, being $\lambda_{\text{peak}}T = 2.898 \times 10^{-3}\,\text{m K}$. It should also be noted that for the temperatures selected in the figure a great deal of energy is radiated outside the visible region.

2.10. The Doppler Effect

Here we are considering a physical phenomenon discovered by the Austrian physicist Christian Doppler (1803–1853; Figure 2.10). This phenomenon will be well-known to those who have attended a Formula 1 race. Situated in the middle of a straight section of track, in which the speeds of the cars are more-or-less constant, it will be noticed that the high-pitched engine noise that is heard as a car approaches suddenly drops in frequency as the car passes and begins to recede. This is the *Doppler Effect* that describes what is detected when a source of constant frequency moves either away from or towards an observer.

A sound wave, which is a vibration of the medium in which it travels, is characterized by three quantities, its speed, V, frequency, ν, and wavelength, λ. A sound is heard because pressure waves

Figure 2.10. Christian Doppler.

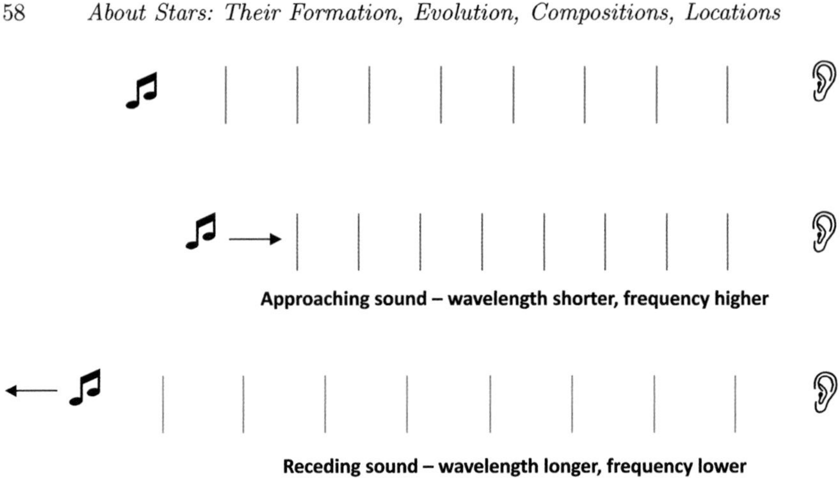

Figure 2.11. An illustration of the Doppler Effect.

impact on detectors within the ear ν times per second and the distance between the pressure peaks as they travel towards the ear is λ. Each pressure maximum travels a distance λ in a time $1/\nu$ and hence the speed of the sound is distance/time or $V = \lambda\nu$. The speed of sound is independent of frequency so that if the wavelength increases the frequency decreases. The law that Doppler found for the variation of wavelength (and hence frequency) with radial speed, v, is

$$\frac{d\lambda}{\lambda} = \frac{v}{V}. \tag{2.40}$$

In (2.40) λ is the wavelength for the source of sound at rest relative to the observer and $d\lambda$, the change of wavelength, is positive with the source moving away from the observer. A representation of the Doppler Effect is given in Figure 2.11; The pressure maxima are drawn apart when the distance between source and observer is increasing and pushed together when the source-observer distance is decreasing.

What is true for sound is also true for light, which is another type of energy propagated in the form of waves. While many of the light sources for which the Doppler Effect is found have radial speeds much less that the speed of light, so that equation (2.40) is applicable, this is not true for distant galaxies. In that case a relativistic Doppler-shift equation must be used, as given in Appendix B.

Problems 2

2.1 Find the gravitational potential energy of a spherical body of radius R within which the density varies as $\rho(x) = a(1-x/R)$.in the form $\alpha GM^2/R$.

2.2 A spherical body of radius 10^8 m consists of gas for which the mean molecular mass is 4×10^{-27} kg. The density can be approximated as $10^8/r$ kg m^{-3} where r is the distance from the centre in metres. What is the average temperature of the gas?

2.3 **(a)** A newly-formed protostar (a star at the beginning of its existence) is in the form of a uniform gas sphere of radius 2×10^{14}m, temperature 20 K and mass 2×10^{30} kg. What is the minimum mean molecular mass of its material if it is to collapse to form a star?

(b) If the protostar had mass 2.4×10^{30} kg, with all other parameters unchanged, then estimate the rate of collapse of its surface, starting from rest, after 1,000 years in metres per second. You may assume that $\ddot{I}$ stays at its initial value during the 1,000 year period.

2.4 **(a)** Assuming that pressure forces are not acting, find the free-fall time for the complete collapse of the protostar described in Problem 2.3(b).

(b) Find the rate of collapse when the radius has fallen to 0.75 of the original radius.

2.5 A gaseous filament, the material of which has mean molecular weight 4×10^{-27} kg, temperature 30 K and density 10^{-12} kg m^{-3}, has a circular cross section of radius 10^{13} m. What are the masses of the condensations formed? Will they be more massive than the Jeans critical mass?

2.6 The Maxwell–Boltzmann distribution is hump-shaped. Find the value of the kinetic energy at the peak of the distribution in terms of the product kT.

2.7 A gas cloud with density $10^{-7}\,\mathrm{kg\,m^{-3}}$ has C^+ as the sole cooling agent, with the cooling rate given by equation (2.35). Find the cooling rates for temperatures 50 K by steps of 100 K to 650 K and plot cooling rates as a function of temperature. Heating, mainly by cosmic rays, is at a rate $4 \times 10^5\,\mathrm{W\,kg^{-1}}$. Find the temperatures at which the cloud is in thermal equilibrium. One of these temperatures represents stable equilibrium and the other unstable equilibrium. Explain why this is so.

2.8 Find the optical depth of the following spherical interstellar clouds and comment on whether heat from a source at the centre will or will not escape on a short timescale.

(a) Radius 20 ly, density $10^{-12}\,\mathrm{kg\,m^{-3}}$, temperature 10 K,
(b) Radius 10^3 km, density $10^{-8}\,\mathrm{kg\,m^{-3}}$, temperature 100 K (Use Figure 2.7 to estimate opacities).

2.9 At what temperatures are the peak wavelengths of the light radiation curves equal to the limits of visible radiation — 400 nm and 700 nm?

2.10 A very prominent line in the spectrum of sodium has a wavelength of 588.9950 nm. When a distant star is observed the wavelength measured is 589.1053 nm. What is the radial speed of the star with respect to the Earth?

(The speed of light is $2.99792 \times 10^8\,\mathrm{m\,s^{-1}}$).

Chapter 3

The Evolution of the Universe

3.1. The Structure of the Universe

So far we have followed the development of the Universe to the stage where it is rapidly expanding, the temperature has fallen to about 60,000 K and atoms of hydrogen, deuterium, helium and a little lithium have formed. There are no discernible dense structures in the Universe and it will have to undergo many changes to become the Universe we have today.

The present contents of the Universe show a hierarchical structure with entities of ever decreasing mass, as seen in Figure 3.1, and it is tempting to believe that this indicates the time-sequence of formation, with the largest structure produced first and then, by repeated subdivision, smaller structures produced in sequence. The elements of the hierarchical structure of interest for star formation are indicated in Figure 3.1 in bold type.

The first entry in Figure 3.1 that most readers will be familiar with is 'Galaxies'. The Solar System is situated just over one-half of the way out from the centre of the Milky Way, a *spiral galaxy* (Figure 3.2(a)) containing about two hundred billion stars. It is estimated that there are about 10^{11} galaxies in the Universe, although they are not all similar to the Milky Way. There are smaller and larger galaxies and many are *elliptical galaxies* (Figure 3.2(b)), elliptical in shape and without any protuberances. The galaxies are not scattered randomly in the Universe but exist in clusters,

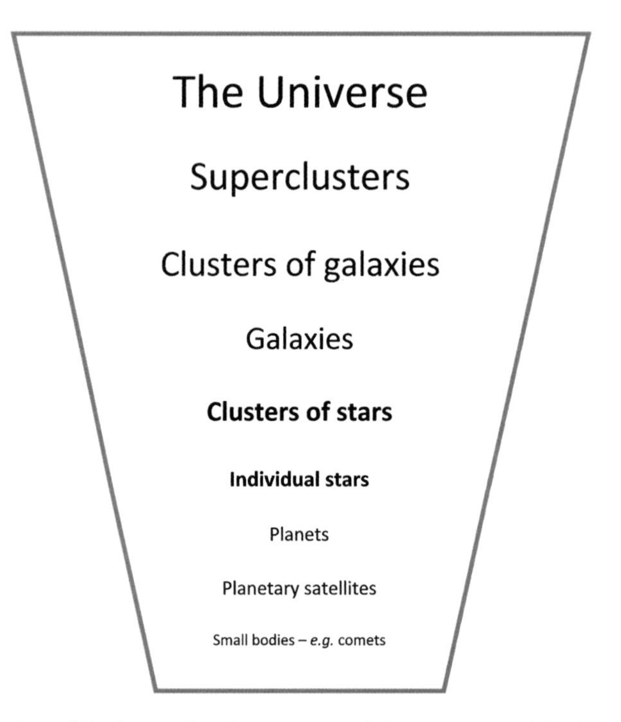

Figure 3.1. The hierarchical structure of the contents of the Universe.

(a) (b)

Figure 3.2. (a) A spiral galaxy (b) an elliptical galaxy.

separated from other clusters by distances that are much larger than the mean distance between galaxies within a cluster. The Milky Way is the member of a cluster known as the Local Group, containing more than 50 galaxies, mostly small; its diameter is about 3 Mpc.[1] The most massive members of the Local Group are the Andromeda Galaxy, followed by the Milky Way.

Even larger organizational structures are *superclusters*, collections of clusters of galaxies separated from other superclusters by distances that are large compared with the mean distance between the clusters The Local Group is member of the Virgo supercluster, containing more than 100 clusters and has a diameter of about 30 Mpc. It has been suggested that even larger structures exist — clusters of superclusters — but this is uncertain.

3.2. The First Condensations

The phenomenon of gravitational instability was described in Section 2.5, whereby a cloud of gas could fragment into a number of condensations with masses of the same order as the Jeans critical mass (Section 2.3). If the condensations collapse and increase their density, but remain cool by radiating away the heat generated by the collapse, then, since the Jeans critical mass is reducing, they may reach a point where, in their turn, they break up through gravitational instability. So, the question is 'did a large body of gas with mass equal to that of a supercluster separate out from Universe material and then in, two stages of gravitational instability, form first separated material that would form clusters of galaxies and by a similar process separated material forming individual galaxies?' The answer to the question is 'no', it did not happen that way.

The reason for rejecting that idea is based on an argument involving the Jeans critical mass. Models of the evolution of the Universe indicate how the mean density and temperature varied with time in the early Universe, as shown in Figure 3.3. One million years after the

[1]Mpc is one million parsecs. A parsec is an astronomical unit of distance that will be defined in Chapter 7. It is equivalent to 3.26 ly or 3.086×10^{16} m.

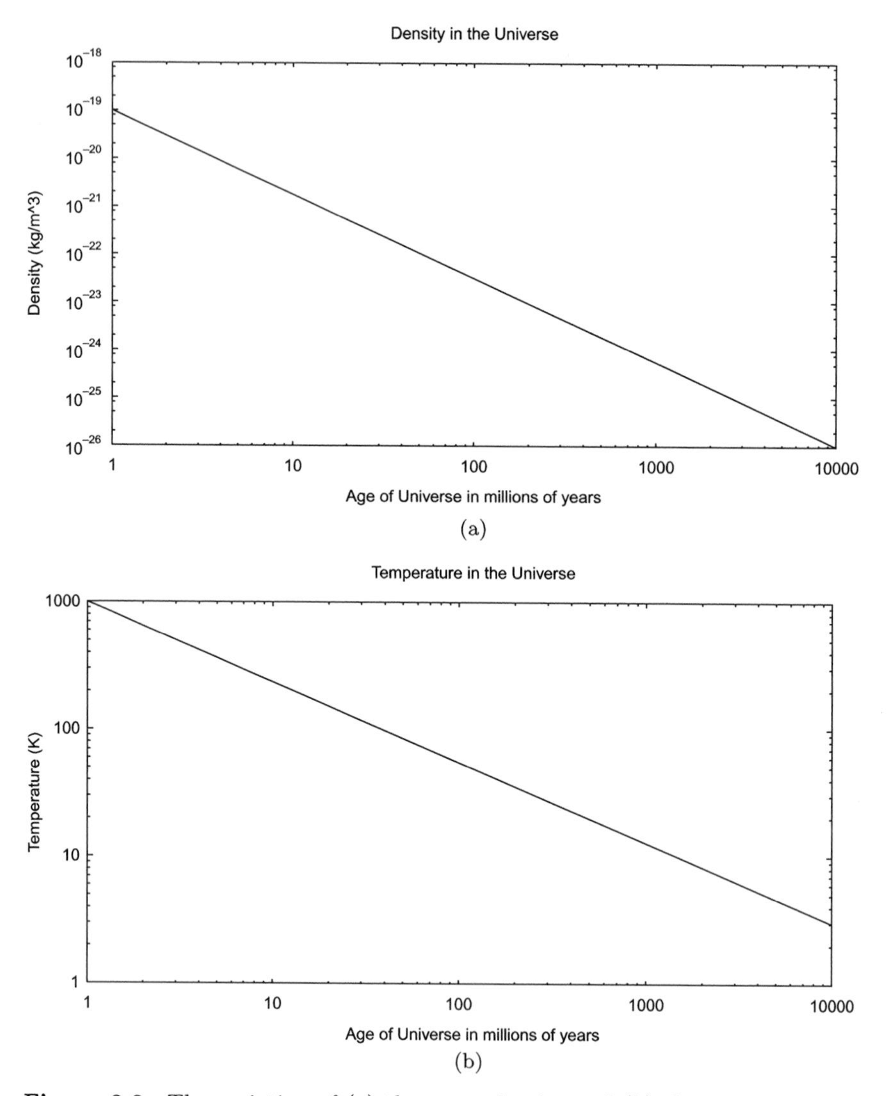

Figure 3.3. The variation of (a) the mean density and (b) the temperature of the Universe with its age.

Big Bang, the density was $10^{-19}\,\mathrm{kg\,m^{-3}}$ and the temperature about 1,000 K. This combination for the material comprising the Universe, mostly hydrogen, corresponds to a Jeans critical mass of 6×10^{35} kg, a considerable mass, the equivalent of 300,000 solar-mass stars, but

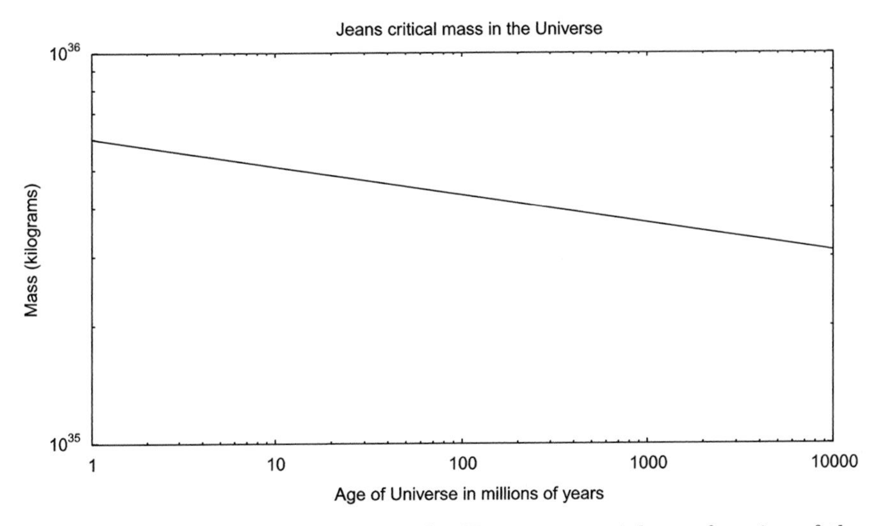

Figure 3.4. The Jeans critical mass for Universe material as a function of the age of the Universe.

less than one-hundred-thousandth of the mass of an average galaxy. Subsequently, the density falls, which increases the Jeans critical mass, and the temperature falls, which decreases it. The net result is that the Jeans critical mass falls with time, but not excessively, as shown in Figure 3.4.

For a condensation to form by gravitational instability it must not be appreciably disturbed during the process of condensation, otherwise the material will be stirred up and dispersed. When the Universe had an age of one million years the free-fall time corresponding to its density was over six million years so the conditions should be not too turbulent over some large fraction of that period. Certainly by one million years the expansion would have considerably slowed down so that local conditions might have been reasonably quiescent.

It appears from the above arguments that the first condensations to form in the expanding Universe contained a few hundred thousand times the mass of the Sun — similar to the mass of a stellar cluster of a type known as a globular cluster. Figure 3.5 shows the globular cluster M13, also known as *The Great Globular Cluster in Hercules.*

Figure 3.5. The globular cluster M13.

It contains about 300,000 stars, has radius 28 pc and is at a distance of 6.8 kpc from the Sun.

We have found that, in the evolution of the Universe, the first condensations to be produced have the masses of globular clusters, entities in the middle of the hierarchical sequence of objects displayed in Figure 3.1. To explain the whole of the figure that is where we must start and develop larger structures going upwards and smaller structures going downward.

3.3. The Development of Galaxies and Larger Structures

The way in which a collapsing cloud of globular-cluster mass could progress to become an actual globular cluster containing stars will be discussed in Section 3.4. For now we shall assume that globular clusters exist and consider how galaxies and larger structures might form.

There are at least two possible ways in which galaxies could form. The first way begins with the Universe populated by a large number of globular clusters, each containing hundreds of thousands

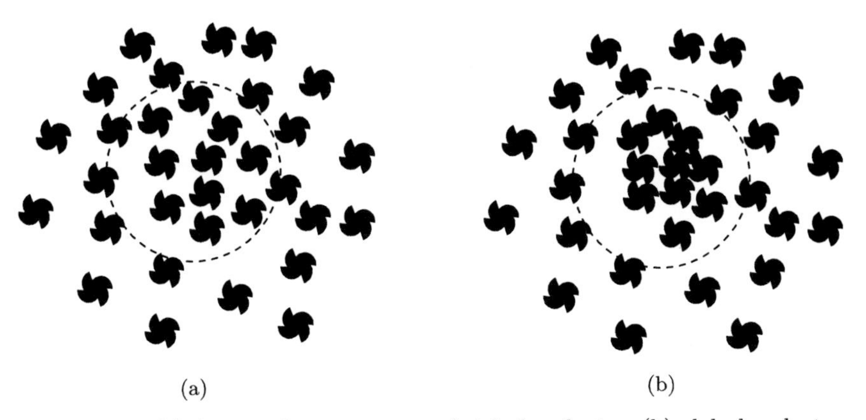

Figure 3.6. (a) An initial arrangement of globular clusters (b) globular clusters aggregating to form a galaxy.

of stars. This is illustrated on a small scale in Figure 3.6(a) that shows an initial distribution of globular clusters. There is a slightly higher density of clusters within the dashed circle so clusters in that vicinity all have a net attraction towards its centre. The effect of this is shown in Figure 3.6(b) where clusters have come together; if the total number of globular clusters coming together is of the order of one hundred thousand then a galaxy would form. This is a process of gravitational instability where even a small excess of the density of globular clusters in one region would amplify with time and precipitate their amalgamation. Then, with galaxies formed in this way, the same kind of process, on larger scales, could lead first to clusters of galaxies and then to superclusters.

The second possible route for galaxy formation that has been suggested is through lumpiness in the early Universe with lumps identified as individual galaxies. The lumps would have been moved apart by the overall expansion of the Universe but the lumps themselves, because they are collapsing due to self-gravitational forces, would not have expanded in size at the same proportional rate. In this way the lumps became increasingly identifiable as separate entities (Figure 3.7). Depending on the scale of the lumpiness each lump could produce a galaxy, a cluster of galaxies or even a supercluster. Whatever kind of entity is first produced by such a process, structures which are higher in the hierarchical sequence in Figure 3.1 would

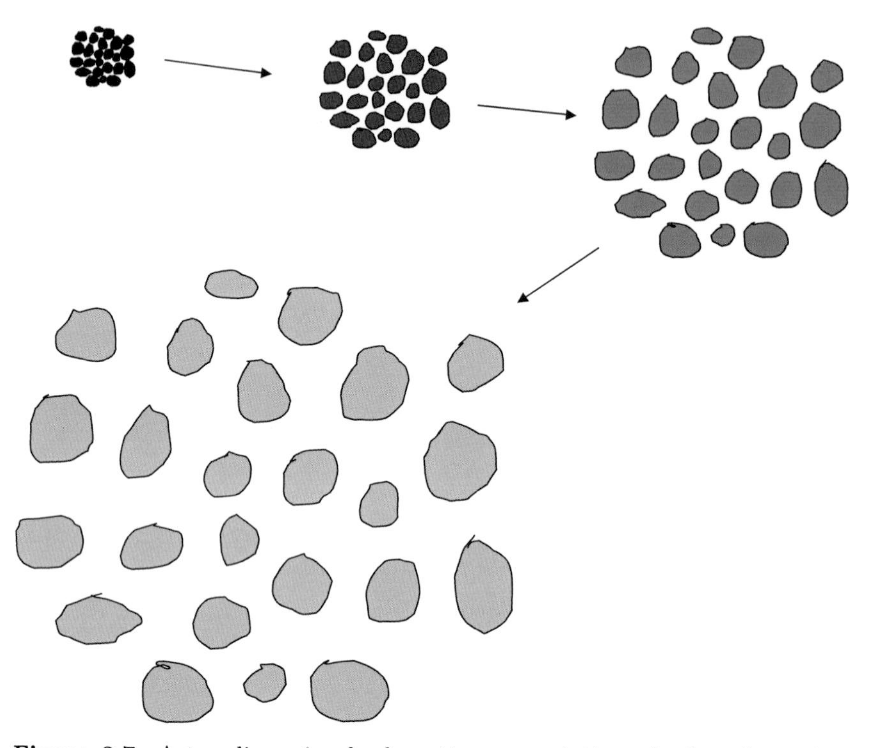

Figure 3.7. A two-dimensional schematic representation of galaxy formation. Overall expansion of the Universe was modified within galactic lumps by gravitational attraction, giving increasing separation of galaxies as the expansion proceeded.

come about by gravitational instability of the type illustrated in Figure 3.6 whereby entities aggregated to form larger entities well separated from other entities of the same kind. Structures lower in the sequence would require some mechanism for breaking a larger structure up into smaller substructures — i.e. a mass of gas corresponding to a globular cluster would have to break up into individual stars, which stayed together to form the globular cluster.

3.4. Forming the First Stars

We start with a collapsing cloud of Universe material with total mass similar to that of a globular cluster. Because it has greater than a Jeans critical mass, corresponding to its material, density and temperature, it has begun slowly to collapse. The gas is increasing in density — essentially it is being compressed — and consequently, if there were no cooling processes available, it would heat up. This is a phenomenon well known to those who ride bicycles; compressing the air in the barrel of a bicycle pump when inflating tyres can make the barrel very hot. However, if the material is very diffuse and transparent to radiation, so that the optical depth of the cloud is small, then the cooling processes described in Section 2.7 will be very effective. In that case the initial stages of collapse will take place with no change of temperature, which will remain at, or close to, that prevailing in the Universe at that time.

Globular clusters, which we take as the earliest dense structures produced in the Universe, contain stars with solar mass and below. To produce these stars we are looking for conditions that can produce a sequence of sub-condensations, the last of which will have a Jeans mass equal to one solar mass or less, and also have a very small optical depth, much less than unity, so they maintain the current temperature of the Universe. First, as a demonstration of the principles involved, we test whether the conditions are suitable for producing a one solar-mass star one million years after the formation of the Universe when the prevailing temperature was 1,000 K (Figure 3.3(b)). From equation (2.16) with the suggested value of Z the required density is

$$\rho = 4.0 \times 10^{42} \frac{T^3}{M_\odot^2} = 10^{-9}\,\mathrm{kg\,m^{-3}}. \tag{3.1}$$

With this density the radius for a solar-mass star is 7.8×10^{12} m. The value of κ for 1,000 K is about 0.22 so the optical depth estimate is

$$\tau = \kappa\rho R = 0.22 \times 10^{-9} \times 7.8 \times 10^{12} = 1.7 \times 10^3.$$

A solar-mass condensation in this state would be almost completely opaque and could not begin an isothermal collapse at this stage of the evolution of the Universe. We now take a more general approach.

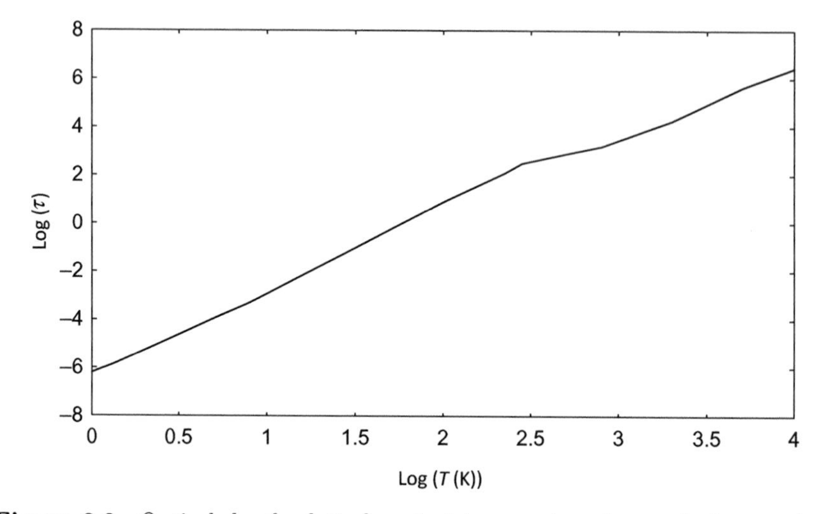

Figure 3.8. Optical depth plotted against temperature for producing a solar-mass star.

For any particular density $R = \left(\frac{3M}{4\pi\rho}\right)^{1/3}$ and using (2.16) to eliminate ρ and with $M = 2.0 \times 10^{30}$ kg,

$$\tau = \kappa\rho R = \kappa\left(\frac{3M\,\rho^2}{4\pi}\right)^{1/3} = \left(\frac{12}{\pi}\right)^{1/3} \times 10^{28}$$

$$\times \frac{\kappa T^2}{M} = 7.8 \times 10^{-3}\kappa T^2. \tag{3.2}$$

The value of $\log(\tau)$ is plotted against $\log(T)$ in Figure 3.8 and it will be seen that for small values of τ the temperature has to be at most a few tens kelvin. From Figure 3.3(a) this suggests globular cluster formation somewhere in the region of 100 million to one billion years after the Big Bang. This is consistent with current age estimates for globular clusters (Section 4.5).

What we have considered so far are the conditions for forming a transparent mass of gas for which the Jeans critical mass is a solar mass. This will eventually collapse to form a star like the Sun. A diffuse body in this state is called a *protostar*. For a temperature of 20 K the protostar will have density $8 \times 10^{-15}\,\mathrm{kg\,m^{-3}}$

and radius 3.91×10^{14} m. Here we introduce a new unit of distance, the *astronomical unit* (au). It is the average Earth-Sun distance and is $1,496 \times 10^{11}$ m. This gives the radius of the protostar as 2,600 au. How this collapses to form a star like the Sun will be described in the following chapter.

Problem 3

3.1 A spherical gaseous condensation in the Early Universe had a mass of 5×10^{35} kg. temperature 500 K. density 10^{-21} and mean molecular mass 4×10^{-27} kg.

(a) What is the Jeans critical mass for this material? (You may take $Z = 2.0 \times 10^{21}$ in equation (2.16)).

(b) If the cloud is initially static then what is the speed of the boundary material after one million years?

Chapter 4

The Formation and Evolution of Stars

4.1. From Protostar to Main-Sequence Star

Chushiro Hayashi (1966) — previously mentioned in relation to radiative cooling (Section 2.7) — described the journey of a protostar of solar mass to the state when it is like the Sun, a *main-sequence star*. His description is best followed on what is known as a *Hertzsprung–Russell (H–R) diagram* (Figure 4.1), one version of which plots the temperature of the evolving star on the x-axis and its *luminosity*, i.e. the total energy of radiation emitted per unit time, on the y-axis. The luminosity is given in units of the solar luminosity, 3.828×10^{26} W.[1] Because the ranges of temperature and luminosity to be covered are so large, the plots are logarithmic, so that 2 on the luminosity scale means that the power radiated is 10^2 times that of the Sun. Similarly 3 on the x-axis indicates a temperature of 10^3 K.

There is a relationship linking the luminosity, L, temperature, T, and radius, R, of a star. The surface area of the star is $4\pi R^2$ and in Section 2.9 it was given that the energy emitted per unit area is

[1]The watt (W) is a scientific unit of power (energy per unit time) and is familiar in the home as indicating the power input of devices such as an electric fire or lamp.

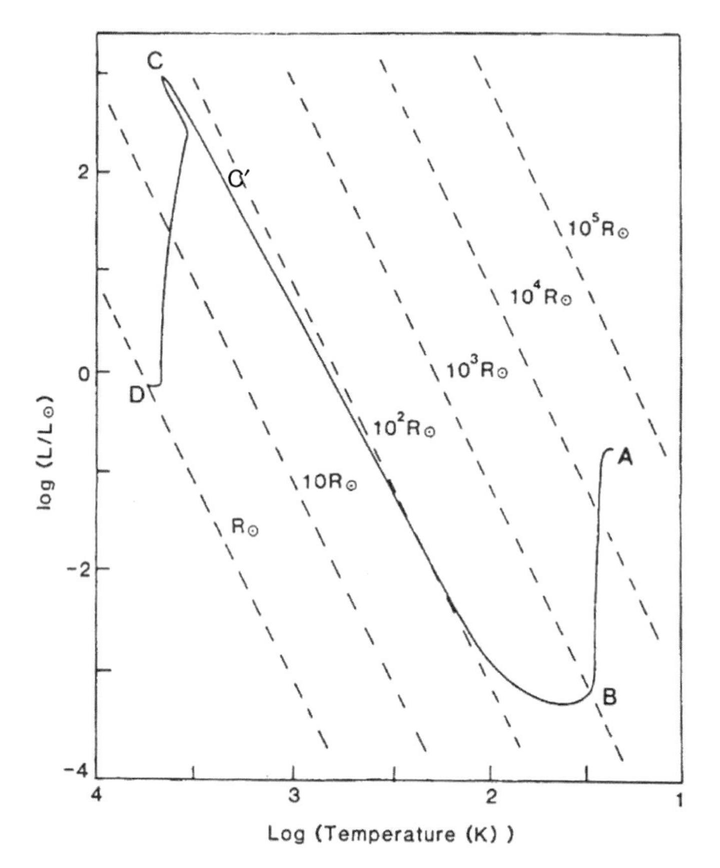

Figure 4.1. The Hayashi plot for a star of one solar mass. $R_\odot$ is the radius of the Sun.

proportional to T^4. This gives the relationship

$$L = 4\pi\sigma R^2 T^4. \qquad (4.1)$$

The constant σ is *Stefan's constant*, $5.67 \times 10^{-8}\,\mathrm{W\,m^{-2}\,K^{-4}}$. Since each point on the H–R diagram indicates luminosity and temperature, the radius at the point is also determined from equation (4.1). The diagram shows lines of constant radius, in units of the solar radius, $R_\odot = 6.957 \times 10^5\,\mathrm{km}$.

The starting point A corresponds to a temperature of 20 K and a luminosity about 20% that of the Sun. Although the radiated power per unit area is very small at 20 K, the comparatively high luminosity is due to the large area of the protostar. Actually the radius of the protostar at A that was taken by Hayashi is only 150 au, corresponding either to one that was unusually small at birth or one that had been collapsing for some time. The journey from A to B involves a large fall in radius but only a small rise in temperature. This is because throughout that period the protostar is transparent to radiation so that heat energy generated by compression of the material is radiated away. At B the density, and optical depth, of the star has increased to a level such that much of the heat generated by the collapse is retained. The increased temperature increases the pressure, which slows down the rate of collapse. The collapse continues with greatly increasing temperature and luminosity to C when a bounce occurs as the protostar moves through an equilibrium position and then returns to it. Hayashi estimated the duration of the bounce as about 100 days. In 1936 there was a temporary increase of the luminosity of a young star, FU Orionis, by a factor of over 200 during a period of less than one year; which could have been a manifestation of the bounce predicted in the Hayashi plot. At point C′, after the bounce, the radius of the star is approximately 50 solar radii, corresponding to a density of $0.01\,\mathrm{kg\,m^{-3}}$, with temperature 4,000 K and luminosity 1,000 times that of the Sun.

The star is now in *quasi-equilibrium*, meaning that at any instant the inward and outward forces on any element of the star are in balance. However, the star is radiating energy so that the equilibrium is being disturbed and the star is continuously changing its configuration — hence the term 'quasi-equilibrium'. The form of evolution from C′ is counter-intuitive. The star shrinks but, despite the fact that it is losing energy by radiation it gets *hotter*! The reason for this is that the energy released by the collapse of the star is sufficient both to increase its temperature and to provide the loss of energy due to radiation. During the journey from C′, known as the *Kelvin–Helmholtz path*, the star is referred to as a *Young stellar object* (YSO). The temperatures indicated in the Hayashi plot are the photosphere temperatures of the star — its effective temperature as a radiator of energy — but internal temperatures are much higher.

When the internal temperature reaches about three million degrees, nuclear reactions involving the hydrogen isotope, deuterium, take place, Deuterium reactions generate too little energy substantially to affect the progress along the Kelvin–Helmholtz path. Eventually, at point D, the central temperature reaches 15 million degrees and then the energy generated by nuclear reactions, converting hydrogen to helium (Appendix C), just balances that being radiated from the surface. The contraction comes to a halt and the star has embarked on the *main-sequence* stage of its existence. Calculated paths of Kelvin–Helmholtz contraction for stars of different masses are shown in Figure 4.2.

The estimation of lifetimes on the main sequence depends on a number of rather rough approximations. The first of these is that the lifetime depends on the amount of hydrogen, m_H, that will be consumed in the core region up to the time that the star leaves the main sequence and that m_H is proportional to the mass of the star, $M*$. This can be expressed as

$$m_H = CM_*, \tag{4.2}$$

where C is a constant.

Empirically it is found that the luminosity of a star, L_*, is related to its mass by

$$L_* = KM_*^{7/2}, \tag{4.3}$$

where K is a constant. Actually the power relationship varies for different stellar mass ranges but (4.3) gives an approximate fit over the whole range.

The luminosity of a star is proportional to the rate that energy is being generated that, in its turn, is proportional to the rate of consumption of hydrogen in the core. Thus we can write

$$\frac{dm_H}{dt} = AL_* = BM_*^{7/2}, \tag{4.4}$$

where A and B are constants.

The duration of the main sequence, t_{ms}, is now taken as the mass of hydrogen available to be consumed in the core divided by the rate

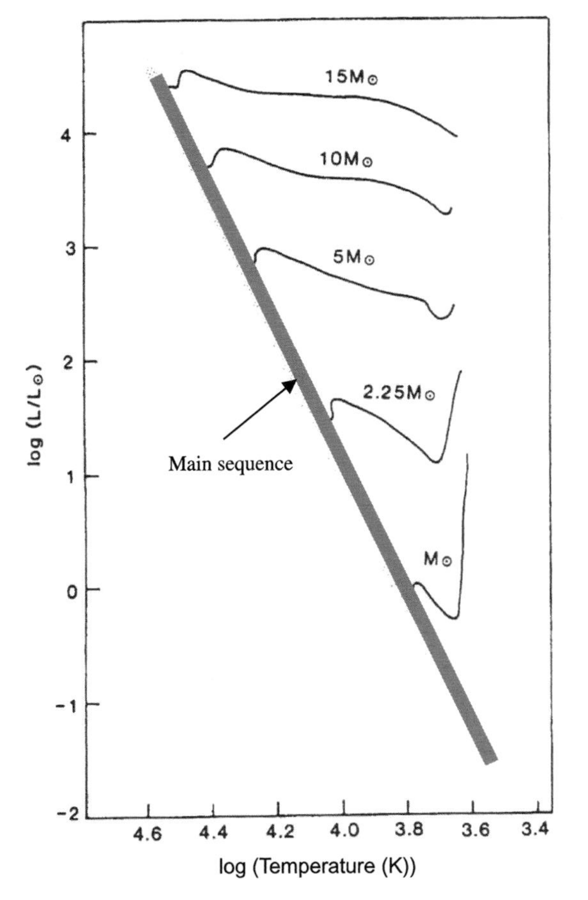

Figure 4.2. Kelvin–Helmholtz paths for various stellar masses. $M_\odot$ is the solar mass.

of consumption so that

$$t_{\text{ms}} = m_H \div \frac{dm_H}{\text{d}t} = CM_* \div BM_*^{7/2} = EM_*^{-5/2}, \qquad (4.5)$$

where E is a further constant. For this relationship to be true the product $t_{ms}M_*^{5/2}$ should be a constant. The actual values for different stellar masses are shown in Table 4.1 together with estimated lifetimes for the Kelvin–Helmholtz stage and for the main sequence. There is a fairly wide variation in the product although, excluding the last entry, it is 43.2 ± 10.8.

Table 4.1. The Kelvin–Helmholtz and main-sequence lifetimes for stars of various masses.

Stellar mass (solar units)	**Kelvin–Helmholtz lifetime (10^6 years)**	**Main-sequence lifetime (10^6 years)**	$t_{ms}M_*^{5/2}$ (10^6 years $M_\odot^{5/2}$)
15	0.062	10	54.0
9	0.15	25	36.5
5	0.58	100	32.4
3	2.5	350	39.0
2.25	5.9	900	44.8
1.5	18	2 000	50.7
1.25	29	4 000	49.6
1	50	10 000	50.0
0.5	150	50 000	26.5

The Sun has been a main-sequence star for about 5 billion years and will continue to be one for another 5 billion years.

4.2. Types of Pressure Within a Star

Once a star leaves the main sequence it will undergo a series of changes of configuration under the influence of pressure forces being generated within it. A useful way to consider pressure is that it is proportional to the translational energy density within a material. If we consider normal gas pressure the translational energy is $3kT/2$ per molecule and hence the energy density is

$$E_d = \frac{3kT}{2}\frac{\rho}{\mu}. \tag{4.6}$$

The gas pressure is given by

$$P_{\text{gas}} = \frac{\rho kT}{\mu}, \tag{4.7}$$

which is two-thirds of the translational energy density.

There are two other sources of pressure within a star — *radiation pressure* and *electron degeneracy pressure.* Radiation pressure is given by

$$P_{\text{rad}} = \frac{4\sigma T^4}{3c}, \tag{4.8}$$

in which σ is Stefan's constant, c the speed of light and T the absolute temperature. The derivation of (4.8) is given in Appendix D.

A material goes into a degenerate state when energy generated within it due to quantum-mechanical effects greatly exceeds the energy generated by its temperature. When fermions of one kind — half-spin particles such as protons, neutrons and electrons — exist together within a single physical system then they must all have different quantum states — a situation required by the *Pauli Exclusion Principle.* It turns out, for reasons that are made apparent in Appendix E, where it is shown in an approximate treatment of degeneracy pressure, that it is the lightest fermions, electrons, which provide the pressure. If we consider highly compressed material, the usual astronomical situation, then atomic structure breaks down and material approximates to an intimate mix of protons, neutrons and electrons. The material becomes a single system within which the Exclusion Principle applies so that every particle must be in a different state from that of all similar particles in the system. The form of electron degeneracy pressure is

$$P_{\mathrm{ed}} = \frac{1}{20}\left(\frac{3}{\pi}\right)^{\frac{2}{3}} \frac{h^2}{m_e}\left(\frac{\rho r}{m_{\mathrm{nuc}}}\right)^{\frac{5}{3}}, \tag{4.9}$$

where h is Planck's constant, m_e the mass of the electron, ρ the density of the material, m_{nuc} the average mass of a nucleon and r the ratio of the number of electrons to the number of nucleons.

In Figure 4.3, we show the regions of density-temperature space in which the various forms of pressure are dominant. Degeneracy can be of two kinds, relativistic and non-relativistic depending on how high the electron degeneracy energies are. Appendix E deals with only the non-relativistic case.

4.3. Evolution from the Main Sequence for Moderate and Low-Mass Stars

The hydrogen-to-helium nuclear reactions (Appendix C) take place most rapidly at the centre of the star where the temperature is highest. Paradoxically the reduction of hydrogen in the core of the star leads to an *increased* rate of energy production there. As the

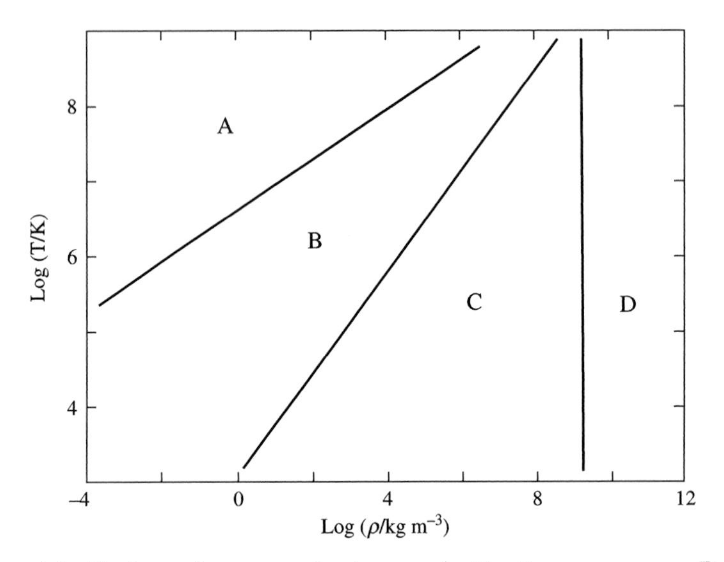

Figure 4.3. Regions of pressure dominance. A. kinetic gas pressure; B. radiation pressure; C. non-relativistic electron degeneracy; D. relativistic electron degeneracy.

hydrogen is exhausted a new state of equilibrium is established by a gradual collapse of the core that increases both its temperature and density and hence the rate at which reactions take place. The increased pressure at the centre, due to the greater generation of energy, expands the star slightly and increases the surface temperature. This part of the evolutionary path for a solar-mass star is from A to B in the H–R diagram, Figure 4.4.

When hydrogen becomes exhausted in the core the surrounding regions, still hydrogen-rich, continues to give reactions in a shell surrounding the core. Progressively hydrogen further out becomes exhausted and shell burning moves outwards. The shell reactions apply pressure both inwards and outwards. Since no reactions are going on in the core it gets steadily compressed to ever-higher densities and its temperature correspondingly increases. The outward pressure expands the star, with the effect that its surface temperature reduces. This is represented in Figure 4.4 by the path from B to C, at which point the star is a *red giant*, with approximate radius 1 au

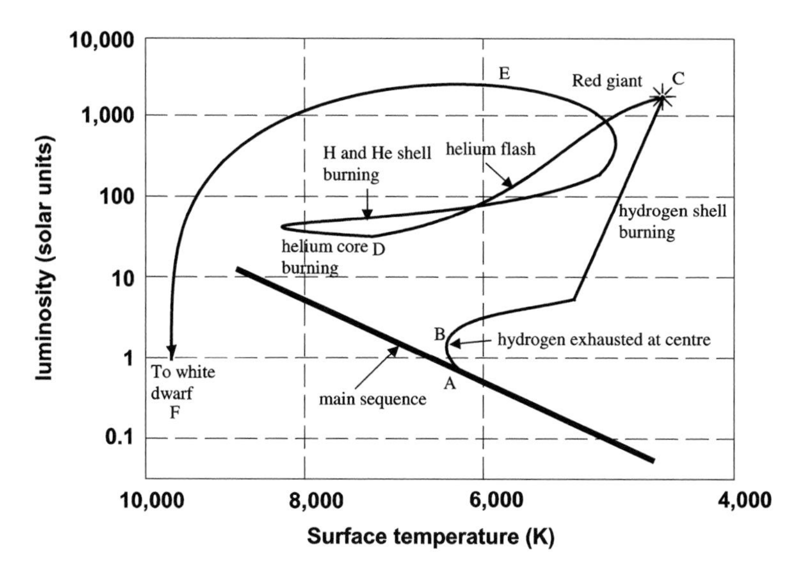

Figure 4.4. The progress of a solar-mass star on leaving the main sequence.

and surface temperature 4,500 K. The Sun will be a red giant in five billion years from now, at which stage it will engulf the terrestrial planets out to Earth.

The core of the star now consists mainly of helium which increases in density and temperature until eventually the temperature reaches about one-hundred million degrees when nuclear reactions can take place in which two helium-4 nuclei combine to produce the isotope beryllium-8 with four protons and four neutrons. However, the only stable beryllium isotope is beryllium-9; some of the beryllium-8, which is unstable with a very short half-life, almost immediately combines with another helium-4 nucleus to give carbon-12 (Figure 4.5). Since the nucleus of a helium-4 atom is an α-particle the set of reactions represented in Figure 4.5 is known as the *triple-αprocess*. The net outcome of these new reactions is that helium-4 is converted into carbon-12.

The helium core, now extremely dense, has become degenerate and only increases its pressure very slightly when the temperature increases greatly (Section 4.2). The triple-α process generates a great deal of heat but, since the pressure increases only by a little, there

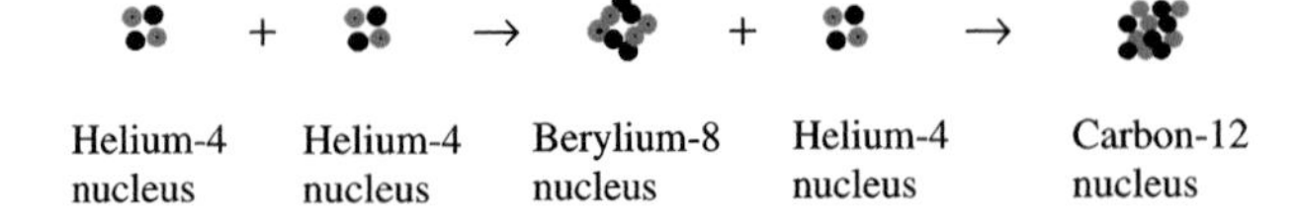

Figure 4.5. The triple-α process.

is no safety-valve mechanism whereby an expansion of the material can reduce both the temperature and pressure. Thus there is a fast and huge temperature increase and the path along CD in Figure 4.4, known as the *helium flash*, which may only last a couple of minutes, gives such a high temperature that the degeneracy of the material is removed and it behaves again like classical material. The star now has helium-core burning, in some ways similar to the original main-sequence state when hydrogen was being consumed in the core. The star evolves towards a configuration resembling the original main sequence — although with a much higher core temperature, larger radius and higher surface temperature.

When the helium in the core is exhausted the core is almost completely carbon — produced by the triple-α process — and there is now helium shell burning around the core and, further out, where the temperature is lower, hydrogen shell burning (Figure 4.6). The star, with its core being compressed and outer regions being expanded by shell burning, evolves towards the region E in Figure 4.4, again resembling a red giant. Because energy generation in the star is so high, the star does not just expand but also expels material violently from its surface. This material is in the form of a complete shell but is seen in projection through a telescope as a ring; it is called a *planetary nebula* (Figure 4.7) although it has nothing to do with planets.

The inward pressure from helium shell burning compresses the core and turns it into a degenerate state once more. The loss of outer material is so great that eventually all that remains of the star is the degenerate core. The star is now a *white dwarf*, corresponding to position F in Figure 4.4, consisting mainly of carbon and with no nuclear reactions taking place. A white dwarf can have the mass of the Sun in a body the size of the Earth, giving a density some two

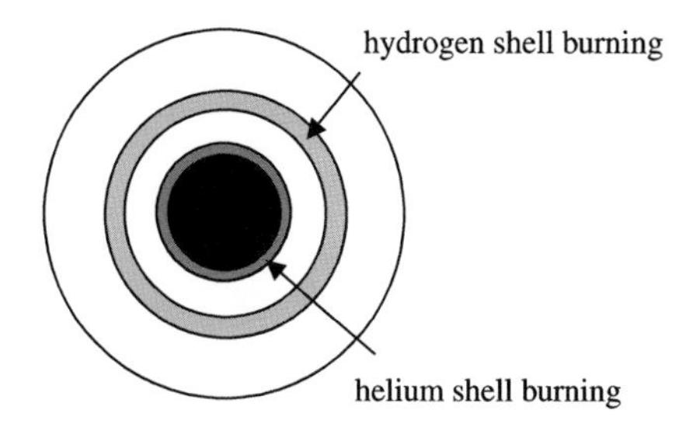

Figure 4.6. Helium and hydrogen shell burning with a carbon core.

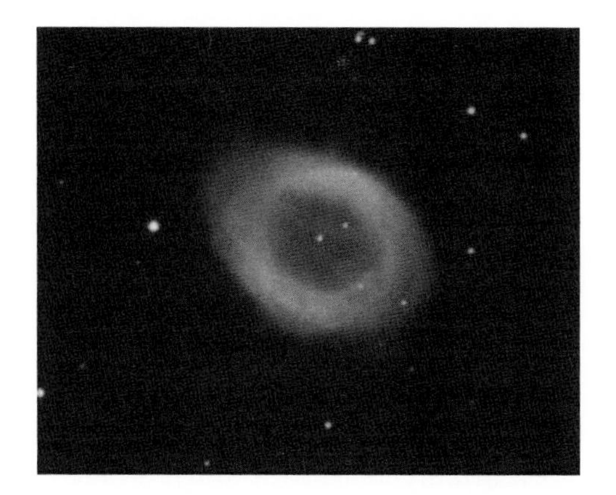

Figure 4.7. A planetary nebula (WIYN/NOAO/NSF).

million times greater than that of water. It shines brightly at first due to the energy stored within it but eventually it cools, emits less and less radiation and ends its existence as a virtually non-radiating body in the state of being a *black dwarf*.

4.4. Evolution from the Main Sequence for High-Mass Stars

The collapse of high-mass stars gives higher temperatures so that triple-α reactions take place in the core before it becomes degenerate. The normal pressure-based safety valve operates and there is no helium flash. Instead ever-heavier nuclei are produced by new reactions such as

Helium-4 + Carbon-12 → Oxygen 16,

Carbon-12 + Carbon-12 → Neon-20 + Helium-4,

Carbon-12 + Carbon-12 → Sodium-23 + Hydrogen-1 (a proton),

Carbon-12 + Carbon-12 → Magnesium-23 + neutron,

Oxygen-16 + Oxygen-16 → Silicon-28 + Helium-4.

Whenever a new type of energy-generating process begins in the core, the star reverts back to a state resembling the main sequence so the path on the H–R diagram oscillates to-and-fro towards the red-giant region then back towards the main sequence region but with ever decreasing amplitudes of swing. The sequence of reactions, giving heavier elements, can progress to the point at which iron is produced and then they can go no further. Reactions up to that stage are *exothermic*, that is they generate energy, heat up the material and so enable new reactions to occur at an increasing rate. Reactions that produce nuclei heavier than iron are *endothermic*, that is they require an input of energy for them to take place. A star with a mass in the range 10–20 $M_\odot$ eventually reaches a stage where there is an iron core with several shells of burning taking place, with different major constituents in each shell, as shown in Figure 4.8.

Now a new process occurs, well outside the range of classical physics. The pressure on the iron core from the external material is so great that the material in the core is unable to support it. The protons and electrons in the core are squeezed together so hard that they combine to form neutrons; now the core consists of closely packed neutrons. The core collapses extremely quickly to a density of $10^{18}\,\mathrm{kg\,m^{-3}}$ and a radius of a few kilometres. Outer material, quickly moving in to fill the space vacated by the original iron core, strikes the neutron core, bounces outwards and collides with material still moving inwards. There is an explosion generating enormous energy — a *supernova*. Figure 4.9 shows the Crab Nebula, debris from a supernova that was observed by Arab and Chinese astronomers in 1054.

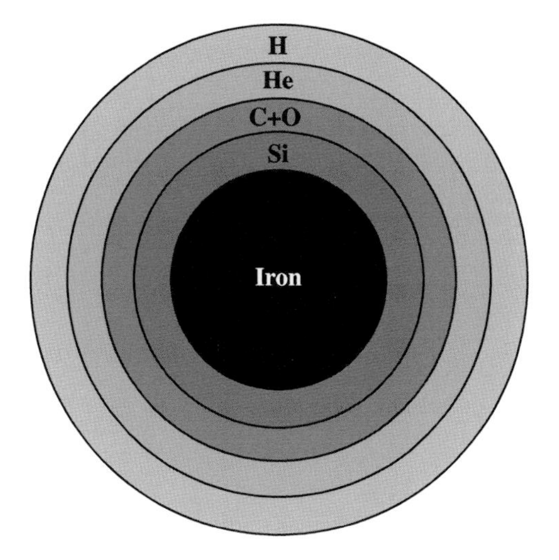

Figure 4.8. Shells in a highly-evolved massive star (not to scale).

Figure 4.9. The Crab Nebula — a supernova remnant (*NASA/ESA*).

The energy generated in a supernova enables endothermic nuclear reactions to take place that produce elements heavier than iron; all such elements that now exist in the Universe are the products of supernovae.

Another product of the Crab supernova is a *neutron star*, a sphere of tightly packed neutrons, a few kilometres in diameter, encapsulated in an iron crust. This is the neutron core of the exploding star. It has been shown theoretically that if the mass of the remnant of the original main-sequence star is greater than about 3–4 $M_\odot$ then gravitational forces on the core would be so great that even neutrons would be crushed. The final outcome in that case is a *black hole*, a mass existing at a point in space that is so massive that not even light can escape from it, so that it can only be detected through its gravitational influence. A black hole is a difficult concept to envisage — but the fields of astronomy and cosmology take us into the behaviour of matter well outside our everyday experience. Later, we shall have more to say about white dwarfs, neutron stars and black holes.

4.5. The Ages of Globular Clusters

Since globular clusters were the first condensed objects to be formed, their ages are of much interest. The oldest globular cluster gives a lower bound to the age of the Universe and, if we accept the age of the Universe from the Hubble plot, it gives an indication of how long after the formation of the Universe globular clusters began to form.

From Table 4.1, it can be seen that Kelvin–Helmholtz lifetimes are much shorter than main-sequence lifetimes so we assume that a star comes into existence at the beginning of its main-sequence stage. In addition the time period within which stars are forming in the cluster is short compared with the main-sequence lifetime of most of the stars, with the exception, perhaps, of very few massive stars. Thus we can assume that at the time the globular cluster formed, or very shortly thereafter, the H–R diagram would have shown a fully occupied main-sequence line, except perhaps at the very top end occupied by very massive stars with short main-sequence lifetimes. However, for the globular cluster 47 Tuscanae the regions occupied by its stars are shown in dark grey in Figure 4.10. The pale grey symbols indicate non-occupied regions of the main sequence.

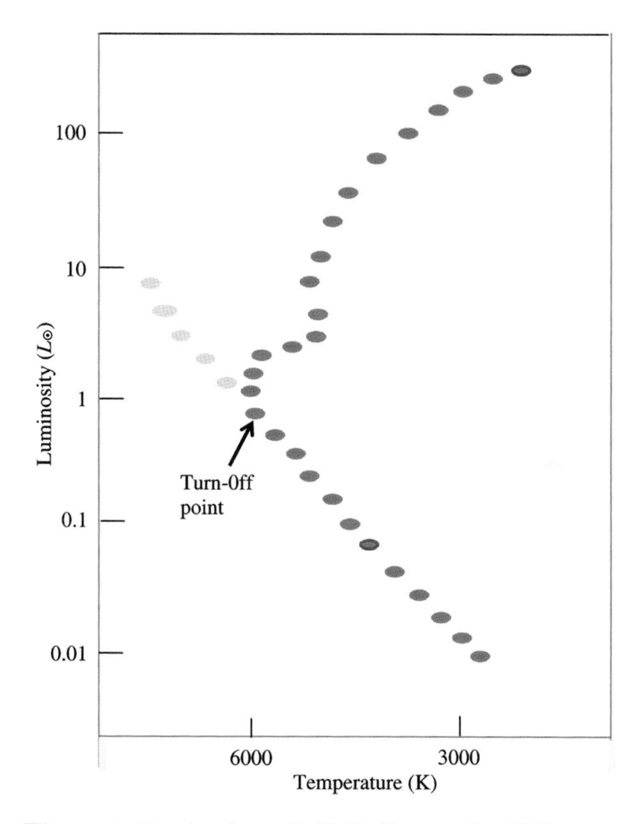

Figure 4.10. A schematic H–R diagram for 47 Tuscana.

From Table 4.1, it is seen that after 100 million years stars more massive than $5\,M_\odot$ would have left the main sequence and, from Figure 4.3, moved off to the right of the main-sequence line. As time progresses so stars of less mass move off the main-sequence line and the point corresponding to the highest-mass star on the main-sequence line is the *turn-off point.* This is an indicator of the age of the globular cluster, essentially the main-sequence lifetime of the most massive stars still on the main sequence. The age estimated for 47 Tuscanae is about 13.1 billion years; most ages of globular clusters are in the region of 13.5 billion years, suggesting that they were formed about 300 million years after the Big Bang. They contain no stars of mass greater than about $0.8\,M_\odot$.

4.6. The Interstellar Medium

Thus far we have been considering the formation of globular clusters, their aggregation to form galaxies, the formation of stars within them and the evolution of those stars. Now we are going forward to the time when our galaxy had formed and consider processes that go on within the galaxy.

The space between stars in our galaxy is not empty but is occupied by very diffuse matter known as the *Interstellar Medium* (ISM). Most of it is hydrogen, about 10% of it is helium and about 1% of it consists of other kinds of element in the form of very fine dust particles. Its density is extremely low, $10^{-21}\,\mathrm{kg\,m^{-3}}$, which means that there is roughly one atom per cubic centimetre. For comparison the air we breathe, consisting mainly of nitrogen and oxygen, contains about 10^{20} atoms in the same volume. The dust particles are about $1\,\mu\mathrm{m}$[2] across and there is about one in every cubic kilometre of the ISM. You might think that the ISM barely exists and can be discounted in astronomical studies but you would be mistaken; the total mass of the ISM is about 5% of that of the stars in our galaxy and everything in the galaxy — stars, planets, satellites, asteroids, comets and you — all originated as ISM material. The ISM is somewhat lumpy with variations of both density and temperature. The average temperature is about 8,000 K, which indicates that the atoms have a high translational kinetic energy corresponding to that temperature. However, because the density of the ISM is so low, its energy density, i.e. energy per unit volume, is also very low and an astronaut within the ISM in an unheated space suit would quickly freeze to death.

4.7. The Formation of Dark Cool Clouds

If we look at a star field through a telescope we see regions where the stars are blotted out by some opaque object in space. This is usually a dark cool cloud (DCC). One such cloud, called the *Horsehead Nebula*, is shown in Figure 4.11. It is part of the *Orion Nebula*, a star-forming region situated in the constellation *Orion*. We now consider how such a cloud can be produced from ISM material.

[2] $1\,\mu\mathrm{m}$ (micron) is 10^{-6} m.

Figure 4.11. The Horsehead Nebula.

ISM material is constantly being heated by the radiation coming from stars and also by *cosmic rays*, which are mostly high-energy particles — almost all hydrogen but with atoms of every other element also present. Unless there were some compensating cooling mechanisms the ISM would have every-increasing temperature, something that does not happen because the cooling mechanisms described in Section 2.7 are also operating.

A supernova not only produces vast amounts of energy but also projects material into a large volume of the surrounding ISM. This has two effects, firstly compressing the ISM by a shock wave and, secondly, injecting material into it, much in the form of fine dust. This increases the cooling of the local ISM due both to dust radiative cooling and cooling by electron collisions. The cooling increases with increasing density and reduces with reducing temperature. With the cooling now exceeding the heating the following sequence of events occurs:

(i) Pressure, P, is proportional to the product ρT, where ρ is the density and T is the absolute temperature. Although ρ has slightly increased the temperature effect dominates and the pressure falls.

(ii) The external ISM pressure is now greater than the local pressure so ISM material enters the cooled region to equalize the pressures. This increases the local density.

(iii) The increased density increases the cooling rate but the lower temperature reduces it. The increased density effect initially dominates so the temperature continues to fall.

(iv) The density continues to rise and the temperature to fall but eventually the two effects balance and the dense region stabilizes.

The outcome is a DCC with higher density than the ISM, normally about $10^{-18}\,\mathrm{kg\,m^{-3}}$, and a lower temperature, in the region 10–20 K. with the DCC in pressure equilibrium with the ISM. This process has been simulated computationally by Golanski and Woolfson (2001).

4.8. The Formation of Protostars

If a DCC has a density and temperature such that its mass is greater than the Jeans critical mass (Section 2.3) then it will begin to collapse. When a DCC collapses, at first the collapse is smooth, with every element of the material moving towards the centre of the cloud. At a certain stage the collapse becomes *turbulent* so that while the general motion is inwards there are streams of gas with components of motion in other directions. Observational evidence exists for turbulence in DCCs. There are sources of microwave radiation in collapsing DCCs, with wavelengths in the centimetre range, coming from transitions between allowed energy states of various substances, such as water and the hydroxyl radical, OH. The laboratory values of the wavelengths are known; from Doppler-shift measurements (Section 2.10) of the wavelengths received on Earth it is deduced that there is turbulence within collapsing DCCs with turbulent streams of dusty gas with speeds up to $20\,\mathrm{km\,s^{-1}}$ (Cook, 1977). An impression of what turbulence is like is shown in Figure 4.12. Anyone who has seen a river passing through a narrow gorge will know that, while the water passes through the gorge in

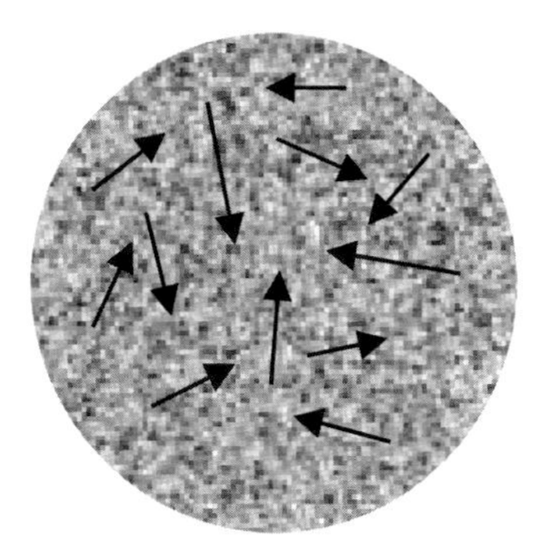

Figure 4.12. A collapsing cloud with turbulence. The general direction of motion of the material (shown by the arrows) is inwards but there is considerable turbulence present.

the general direction of flow, turbulent elements of it are moving in many directions.

Because of turbulence, streams of gas sometimes collide and if the collision is more-or-less head-on then the gas will be compressed and its density increased. When it is compressed it also heats up. For the compressed gas the increased density reduces the Jeans critical mass but the increased temperature raises it. In general, at the time of formation, the compressed material is unlikely to begin to collapse. However, cooling processes, mediated by electrons that have high speeds, are much faster than re-expansion, which depends on the much lower speeds of individual atoms. Consequently the compressed material cools before it has substantially expanded and its mass can become greater than the Jeans critical mass and it can begin to collapse (Woolfson, 1979). The collapsing body can eventually form a protostar, with typical radius 2,000 au and density $10^{-14}\,\mathrm{kg\,m^{-3}}$, giving a protostar mass of $0.57\,M_{\odot}$. We have already described how such a protostar could evolve, firstly they give a main-sequence star and then to leave the main sequence in a way that is mass dependent.

4.9. Types of Clusters and Their Locations

Within a particular star-forming cloud, stars will form in clusters, typically containing a few hundred stars. They are initially held together by the gravitational influence of both the stars and the gas within which they are immersed. In this state the cluster is said to be *embedded*, meaning that the stars are embedded in gas. Eventually the gas is expelled, slowly at first due to the effect of stellar radiation and finally, much more quickly, by the energy generated by supernovae. With the gravitational binding effect of the gas removed most clusters completely disperse to give individual isolated stars or binary systems — two stars linked together — known as *field stars* or *field binaries*. However, in about 10% of cases the cluster retains its cohesion and becomes a *galactic cluster*, sometimes called an *open cluster*. A typical galactic cluster is shown in Figure 4.13.

It has already been indicated that the globular clusters now observed were formed in the early stages of the development of the Universe. From time-to-time a star on the periphery of the cluster will acquire a speed sufficient for it to escape from the cluster, a process akin to the evaporation of liquids. Over the course of time the population of the globular cluster will fall but theoretical calculations

Figure 4.13. The open cluster M45 — Pleiades.

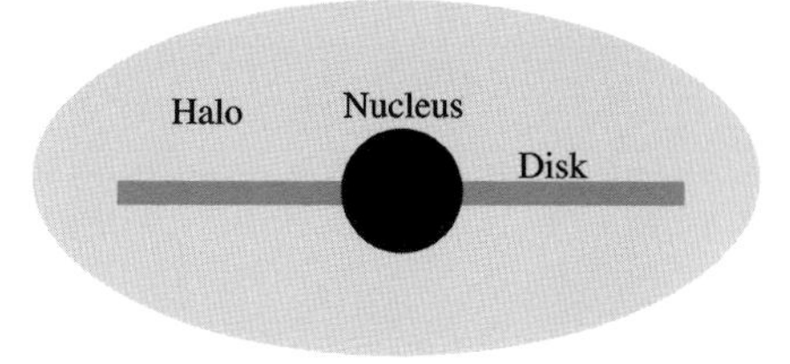

Figure 4.14. A schematic view of the Milky Way galaxy seen edge-on.

show that the prospective lifetime of a globular cluster — the time for it to completely evaporate — is much greater than the age of the Universe. The globular clusters seen today are little different in terms of star numbers from what they were when they first came into existence. By contrast galactic clusters have lifetimes estimated to be in the range 10^8–10^9 years. They are transient objects, constantly disappearing due to evaporation and constantly being formed.

Figure 3.2(a) gives a good impression of the plan view of a spiral galaxy but not of a side projection, shown schematically in Figure 4.14. Open clusters occur only in the mean plane of the galaxy, marked as the disk in the figure. However, globular clusters can occur anywhere — in the disk, the dense nucleus of the galaxy or in the very diffuse halo that envelops the whole galaxy.

Problems 4

4.1 A star is estimated to have luminosity $3 \times 10^{30}\,\mathrm{W}$ and a temperature of 3,900 K. What is its radius?

4.2 The structure of the Milky Way can be approximated as a disk of diameter 10^5 ly and thickness 1,000 ly. It contains about 10^{11} stars of average mass $0.8\,M_\odot$ and interstellar medium of density $10^{-21}\,\mathrm{kg\,m^{-3}}$. What proportion of the total mass of the galaxy is ISM material?

4.3 Calculate the kinetic gas pressure, the radiation pressure and the electron degeneracy pressure for atomic hydrogen under the following conditions:

	$\rho(\mathrm{kg\,m^{-3}})$	$T(\mathrm{K})$
(i)	10^3	10^4
(ii)	10^4	10^7
(iii)	10^4	10^8
(iv)	10^6	10^7
(v)	10^6	10^9

Do not take ionization of hydrogen into account.

Part 3

The Structure and Composition of Stars

Chapter 5

The Equilibrium of Main-Sequence Stars

A main-sequence star is in a condition where it is slowly evolving but always in a state of quasi-equilibrium. Here we shall investigate the equilibrium conditions that must be satisfied, a necessary first step to producing theoretical models of stars.

5.1. Conditions for Modelling a Main-Sequence Star

For producing a model of a star like the Sun it is assumed that the star has spherical symmetry. Because of its spin this would not be strictly true — it would tend to be an oblate spheroid, bulging out at the equator — but since the centripetal acceleration at the equator, $5 \times 10^{-3}\,\mathrm{m\,s^{-2}}$, is much smaller than the acceleration there due to the Sun's mass, $274\,\mathrm{m\,s^{-2}}$, the distortion of the Sun will be negligible, With the assumption of spherical symmetry it is only necessary to find the conditions in a star as a function of the distance from its centre, r. Under the conditions within a star, material behaves like a perfect gas so if the pressure $P(r)$ and temperature $T(r)$ are found then the density is also known if the composition of the material is known.

For stars of solar mass, or a little greater, the normal gas pressure is much greater than radiation pressure or degeneracy pressure so that the last two may be ignored. For conditions deep in the Sun the density will be of order $10^4\,\mathrm{kg\,m^{-3}}$, the mean particle mass $\sim 10^{-27}\,\mathrm{kg}$

(the material is highly ionized) and the temperature $\sim 10^7$ K. Under these conditions the values of P_{gas} and P_{rad} from equations (4.7) and (4.8) are 1.4×10^{15} and $2.5 \times 10^{12}\,\text{N m}^{-2}$, respectively, showing that the radiation pressure can be ignored. However, for more massive stars, where higher temperatures prevail, radiation pressure can be either important or even dominant.

In modelling a star there are two quantities, other than pressure and temperature that are usually found. One is the *included mass*, $M(r)$, the mass of material within the spherical surface of radius r and the luminosity, $L(r)$, the rate of flow of energy across that surface. Two other quantities, related to the physics of the stellar material, are involved in setting up the equations for equilibrium. The first of these is the *intrinsic energy generation function*, ε. At high temperatures nuclear reactions (Appendix C) generate energy at a rate that depends on the composition of the material and its density and temperature. It is expressed in terms of power output per unit mass, or W kg^{-1}. The second quantity is *opacity* that describes the resistance of the material to the flow of radiation through it (Section 2.8).

All the factors involved in determining the basic equilibrium equations for a star have now been described.

5.2. The Pressure Gradient

In Figure 5.1 there is shown a slice of stellar material, of unit area and with surfaces at distance r and $r + dr$ from the centre. The slice is in equilibrium so the total force on it must be zero. There are two pressure forces P and $P + dP$, shown in the figure and also the gravitational force due to the interior material contained within the surface of radius r. Equating the sum of these forces to zero

$$P - (P + dP) - \frac{GM(r)\rho dr}{r^2} = 0,$$

or, with the use of equation (4.7)

$$\frac{dP}{dr} = -\frac{GM(r)\rho}{r^2} = -\frac{GM(r)P\mu}{kTr^2}. \tag{5.1}$$

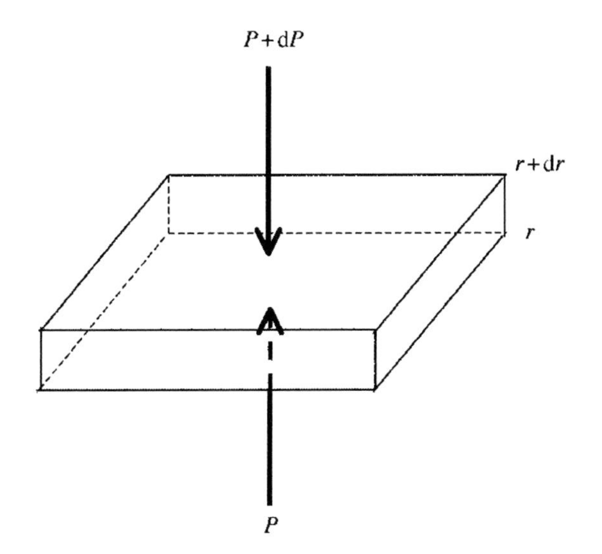

Figure 5.1. The pressure forces on a thin slab of thickness dr.

The boundary condition for pressure is that it must be zero (actually very small) at the boundary of the star where ρ becomes effectively zero (also actually very small).

5.3. The Gradient of Included Mass

If we consider the slice of material shown in Figure 5.1 the increase in included mass between r and $r+dr$, $dM(r)$ is just the mass contained within a shell of radius r and thickness dr. This gives

$$dM(r) = 4\pi r^2 \rho dr,$$

giving

$$\frac{dM(r)}{dr} = 4\pi r^2 \rho = \frac{4\pi r^2 P\mu}{kT}. \tag{5.2}$$

The boundary conditions for $M(r)$ are that $M(0) = 0$ and $M(R) = M_*$ where R is the radius of the star and M_* its mass. For some methods of modelling stars M_* is not a predetermined quantity put into the modelling but rather a quantity that is found as a result of the modelling process.

5.4. The Luminosity Gradient

For equilibrium the energy content of the material between spherical surfaces of radius r and $r + dr$ remains constant. Hence, by energy conservation, the radiation flux through the outer surface exceeds that through the inner surface by the power generated within the shell of thickness dr, which gives

$$L + dL - L = 4\pi r^2 \rho\varepsilon dr,$$

or

$$\frac{dL}{dr} = 4\pi r^2 \rho\varepsilon = \frac{4\pi r^2 P\mu\varepsilon}{kT}. \tag{5.3}$$

The boundary conditions for luminosity are $L(0) = 0$ and $L(R) = L_*$, the luminosity of the star. As for the mass of the star, L_* is normally a derived quantity from the calculations.

5.5. The Temperature Gradient

Figure 5.2 shows a thin slab of material of unit area with faces at r and $r + dr$. Radiation passes through the slab and some radiation

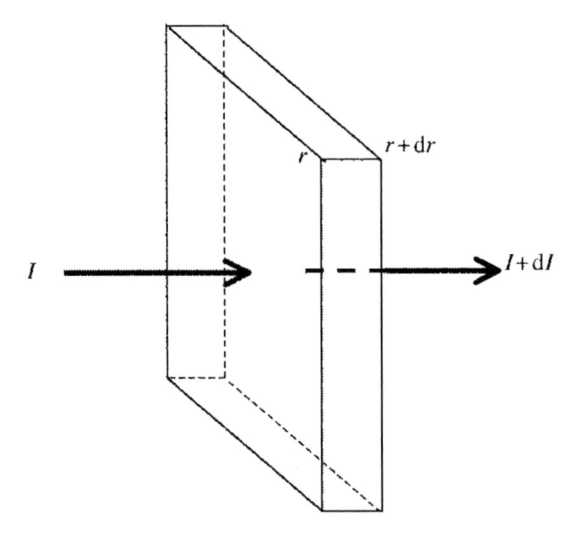

Figure 5.2. The change of intensity of a beam passing through a thin slab of thickness dr.

may be generated within it due to nuclear reactions. With I the *mean* intensity of the radiation within the slab then the energy absorbed per unit area per unit time in passing from the inner to outer surfaces of the slab is

$$\Delta E = I\kappa\rho \; dr.$$

This will be true for a thin slab even if I is being augmented by energy generation during its passage. The intensity of the radiation passing through the slab. i.e. the energy per unit area per unit time, may be expressed in terms of the local luminosity by

$$I = \frac{L}{4\pi r^2}. \tag{5.4}$$

Because of the relationship between energy, E, and momentum, p, for radiation

$$E = pc,$$

the change in momentum of the radiation per unit time between the inner and outer surface is

$$\frac{dp}{dt} = \frac{\Delta E}{c} = \frac{I\kappa\rho}{c} dr = \frac{L\kappa\rho}{4\pi r^2 c} dr. \tag{5.5}$$

This rate of change of momentum in the section of unit area represents a difference in, force per unit area, or radiation pressure between the inner and outer surfaces of the slab, or

$$\frac{dP_{\text{rad}}}{dr} = -\frac{L\kappa\rho}{4\pi r^2 c}. \tag{5.6}$$

The negative sign indicates that the radiation pressure falls with increasing r. Differentiating the expression for P_{rad} in equation (4.8) with respect to r gives

$$\frac{dP_{\text{rad}}}{dr} = \frac{16\sigma T^3}{3c}\frac{dT}{dr}.$$

Equating the two values of dP_{rad}/dr gives

$$\frac{dT}{dr} = -\frac{3L\kappa\rho}{64\pi\sigma T^3 r^2}. \tag{5.7}$$

The boundary condition for temperature is that $T = 0$ at $r = R$. The conventional temperature of a star, as determined by observation, is the temperature of the photosphere, the visible boundary of the star, which corresponds to a density and temperature that are very small compared to the corresponding quantities within the star. For this reason it is usual to take them as zero. The estimates of total mass, radius and luminosity, as found by computation, are insensitive to the actual boundary values taken for ρ and T, as long as they are small compared to interior values.

5.6. Modelling Stars

The four basic equations — (5.1), (5.2), (5.3) and (5.7) provide a set of coupled differential equations that enable the determination of $P(r)$, $M(r)$, $L(r)$ and $T(r)$ throughout a star It has been assumed that there is no convection within the star, where stellar material rises and falls, but convection does occur in most stars so the stellar models will be flawed to some extent. Nevertheless the models indicate the distributions of density and temperature reasonably well.

To do the calculations the composition of the stellar material must be known and also the values of ε and κ as functions of density and temperature either in analytical or tabular form. One method of solution is to fix a radius and to guess the values of P, $M(r)$, L and T at some intermediate value of r, say at $r = {}^1\!/_2 R$. The basic equations are then integrated both inwards towards $r = 0$ and outwards towards $r = R$. In general the values of L and M will not be zero at $r = 0$ and the values of P and T will not be zero at $r = R$. By changing the initially-guessed values by a small amount, one at a time, the rates of change of the boundary values can be found for rates of change of the dependent variables at the intermediate point. In this way better values can be found for the starting point dependent variables and the process is repeated until convergence to an acceptable solution is reached. There are other, and probably better, ways of solving the equations but the simple method described here shows that it is possible in principle.

Problem 5

5.1 At a point within a star the conditions are as follows: density, $\rho = 10^4\,\mathrm{kg\,m^{-3}}$; temperature, $T = 6 \times 10^6\,\mathrm{K}$; opacity, $\kappa = 5\,\mathrm{m^2\,kg^{-1}}$; distance from star centre, $r = 2.2 \times 10^8\,\mathrm{m}$; included mass, $M_I = 1.3 \times 10^{30}\,\mathrm{kg}$; energy production rate, $\varepsilon = 1.5 \times 10^{-4}\,\mathrm{W\,kg^{-1}}$; luminosity, $L = 2 \times 10^{26}\,\mathrm{W}$.

Assuming that the local material, including the effect of ionization, has mean particle mass $10^{-27}\,\mathrm{kg}$, then estimate the pressure at the point. What would be the approximate pressure, included mass, temperature and luminosity at a point 10.000 km closer to the centre?

Chapter 6

Finding the Compositions of Stars

6.1. Atoms, Isotopes, Molecules, Ions and Energy Levels

In Section 1.2, an atom was shown to consist of a nucleus, containing protons and neutrons, which contains virtually all the atomic mass but is very tiny with a diameter of order a few times 10^{-15} m, and a surrounding cloud of electrons, which has very little mass but occupies a space with diameter a few times 10^{-10} m and so gives an atom its overall size. A representation of a carbon atom, with six protons, six neutrons and six electrons, is shown in Figure 6.1. We represent this as C-12, where C is the chemical symbol for carbon and 12 is the number of nucleons — protons plus neutrons. There is another stable form of carbon, C-13 with seven neutrons. C-12 and C-13 are stable isotopes of carbon; it is the number of protons that define the chemical nature of an atom. Six protons in the nucleus can only be carbon and the element represented by C must have six protons.

There are many unstable radioactive isotopes of carbon. The best known is C-14, which has eight neutrons in its nucleus. It has a half-life.[1] of 5,730 years and it is used for dating archaeological artefacts or remains of organic origin.

[1]The half-life of a radioactive isotope is the time taken for one-half of it to decay.

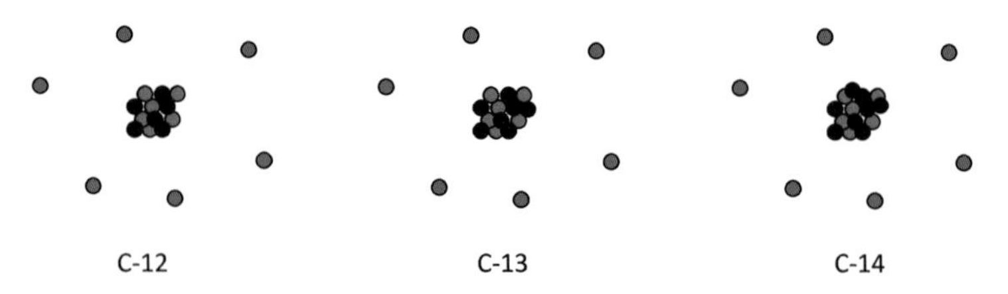

Figure 6.1. A representation of isotopes of carbon (not to scale). C-12 and C-13 are stable and C-14 radioactive.

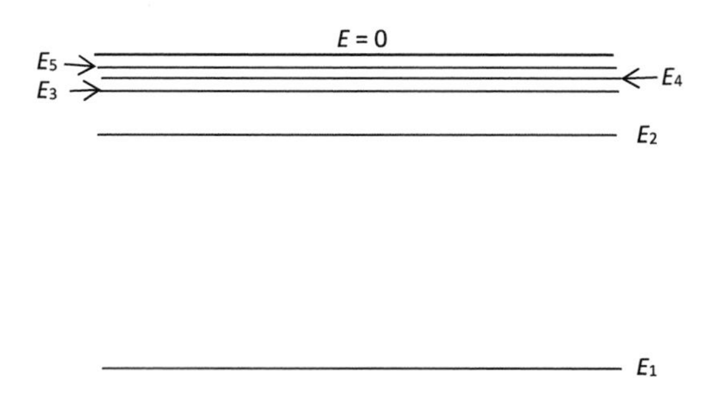

Figure 6.2. Representation of the energy levels in an atom.

A *molecule* consists of a number of atoms bonded together to form a coherent distinctive entity. Two simple examples of molecules with atoms linked by *chemical bonds*, are shown in Figure 1.2.

The configurations of atomic electrons are not randomly related to the nucleus. The branch of physics known as quantum mechanics indicates that the atomic electrons can only have certain allowed negative energies. The fact that they are negative indicates that they are bound to the nucleus; a positive energy would indicate a free electron, one not bound to the nucleus. A set of energy levels are shown in Figure 6.2 and an electron cannot have an energy between these levels.

There are also energy levels for molecules, where the molecule can only adopt specific conformations allowed by the requirements of quantum mechanics.

If an atomic electron with negative energy — $|E|$ somehow has its energy increased by $|E|$ or more, then its energy will become positive

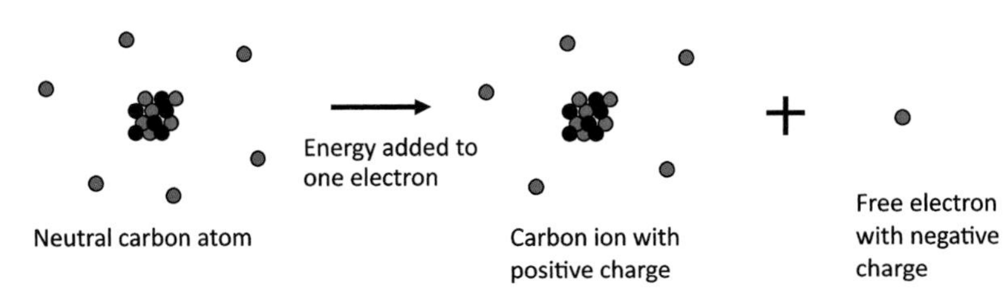

Figure 6.3. Ionization of a carbon atom to give a carbon ion plus a free electron.

and it will escape from the atom, leaving behind an *ion*; it is also possible for the atom to lose more than one electron and be *doubly ionized* or *triply ionized* etc. An example of ionization is illustrated in Figure 6.3 that shows that if energy is added to one electron of a carbon atom, causing it to escape, then the result is singly ionized carbon plus a free electron.

6.2. The Nature of Light

In the 17th century, there was a lively dispute concerning the nature of light. The eminent Dutch scientist and mathematician, Christian Huygens (1629–1695: Figure 6.4(a)) proposed that light was a wave motion, with fluctuations like waves on water, while Isaac Newton (1642–1727; Figure 6.4(b)), probably the greatest scientist of all time, favoured the idea that light was a stream of 'corpuscles' or particles. The dispute seemed to be resolved when the English polymath and scientist, Thomas Young (1773–1829: Figure 6.4(c)) carried out experiments with light showing the phenomenon of diffraction, a characteristic only shown by waves. Figure 6.5 illustrates the difference between the behaviour of a wave motion and of a stream of particles when they meet a gap in a barrier.

At the beginning of the 20th century a phenomenon known as the *photoelectric effect* was raising difficult problems. It was found that if light fell on the surface of some metals, electrons acquired enough energy to escape from the surface. On the basis that light is a wave, its energy would be uniformly spread over the surface of the metal and, depending on the intensity of the light there should be a delay in electron emission while electrons received enough energy

Figure 6.4. The *dramatis personae* in developing an understanding of the nature of light. (a) Christian Huygens, (b) Isaac Newton, (c) Thomas Young, (d) Albert Einstein.

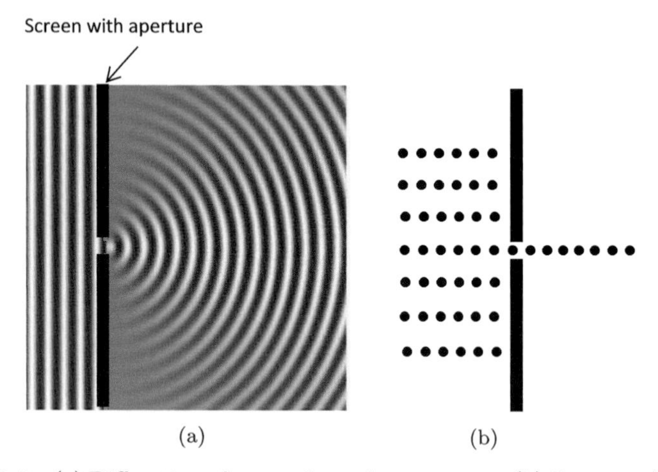

Figure 6.5. (a) Diffraction of waves through an aperture (b) Passage of particles through an aperture.

to escape. Hence there should be a long delay in electron emission for a very low intensity of light but this was not so. Electrons were emitted immediately with the rate of emission proportional to the intensity of the light. Another experimental result was that the shorter was the wavelength of the light — or the higher the frequency — the greater was the energy of the escaping electrons. In 1905 the photoelectric effect was fully explained by the German, later American, scientist Albert Einstein (1879–1955: Figure 6.4(d)). In the photoelectric effect light behaves like a stream of particles, called *photons*, where the energy of a photon is proportional to the frequency, ν, of the light, the relationship being

$$E_{\text{photon}} = h\nu, \tag{6.1}$$

where h is *Planck's constant*..[2] The rate at which photons fell on the metal was proportional to the intensity of the light beam and for a low intensity there would be no delay in emission, just a reduction in the rate of electron emission. The more energetic a photon the greater would be the energy of the emitted electron dislodged by an impacting photon, as is observed.

Einstein's scientific paper on the photoelectric effect was published in 1905 and this work is not as well-known as his theories of relativity. He was awarded the Nobel Prize for Physics in 1921, not for his work on relativity, which was at that time still treated with suspicion by some scientists, but for his work on the photoelectric effect, a lesser achievement but more widely accepted.

The ability of light to display the characteristics of either a wave or a particle, depending on the scenario in which it occurs, is known as *wave–particle duality*. Electrons also have this property; an electron falling on a fluorescent screen gives a point of light and it can be deflected by electric and magnetic fields, showing that it is behaving like a particle; because it has these properties the early television sets using cathode-ray tubes were able to produce images. However, electrons in an electron microscope form an image by behaving like waves.

6.3. Fraunhofer Lines: The Interaction of Light with Atoms

It is well-known that a white-light source, or sunlight, can be spread out into a spectrum by passing it through a prism or a diffraction grating. If this is done for sunlight with high resolution

[2]Planck's constant, $h = 6.636 \times 10^{-34}\,\mathrm{m^2\,kg\,s^{-1}}$.

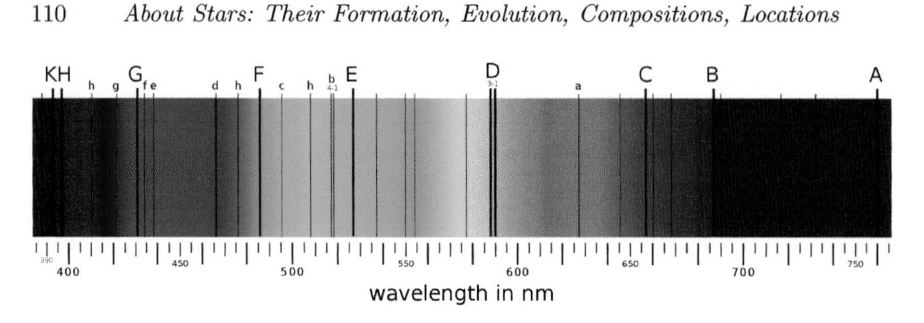

Figure 6.6. A solar spectrum with Fraunhofer lines.

then dark lines will be seen in the spectrum corresponding to missing wavelengths (Figure 6.6); these are known as *Fraunhofer lines*, named after their discoverer, the German physicist Joseph von Fraunhofer (1787–1826). We now describe how these occur.

First we consider what happens when atoms are heated. Some of the heat energy is imparted to atomic electrons and they may go from a lower energy state E_m to a higher energy state E_n, both energies being one of those allowed by the rules of quantum mechanics. Now, a general principle that applies to all physical systems is that, if they can, they go to the lowest energy state possible, which gives them stability. In the case we are considering the electron will jump to a lower energy state, which may be that from which it originally came but could be an intermediate state. Another general rule of physics is that energy must be conserved so, since in falling from energy E_n to, say, energy E_m the electron has lost energy $\Delta E = E_n - E_m$, that energy must be converted into some other form. The form it takes is a photon, with the frequency of the corresponding light comes from the equation

$$h\nu = \Delta E, \tag{6.2}$$

that is clearly related to equation (6.1).

Each type of atom will show characteristic *emission lines* from which the atom can be identified. If a substance containing sodium, say common salt that is sodium chloride, is put in a flame the emission lines shown in Figure 6.7 will be seen; they can be produced by sodium and nothing else.

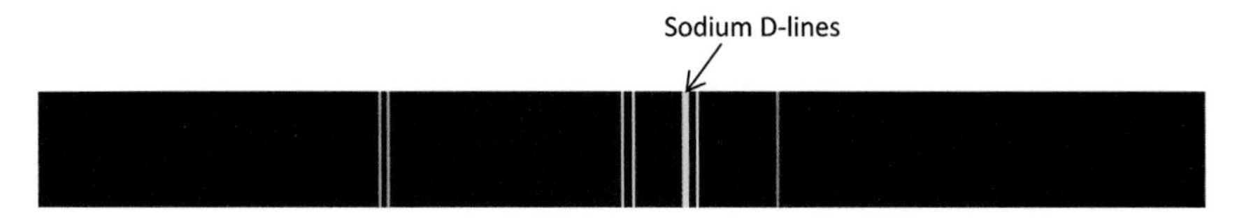

Figure 6.7. The emission spectrum of sodium.

Now we consider what happens to the light coming from a star, which we take as the Sun. The light from the Sun comes from a thin layer called the *photosphere.* Light is generated below the photosphere but the dense solar material absorbs it before it reaches the photosphere. Material above the photosphere is very diffuse and does not emit much light but it does allow the passage of most of the white-light emitted from the photosphere. This diffuse material contains all the elements present in the outer regions of the Sun in vaporized form, including sodium. When a photon from the photosphere has *precisely* the energy corresponding to one of the energy transitions in sodium it is absorbed by a sodium atom, pushing an electron from the lower energy of the transition to the higher energy. As we previously noted, it will revert to a lower energy state again, emitting a photon with a frequency corresponding to one of the emission lines of sodium. However, this photon is emitted in a random direction and so is unlikely to be moving in the direction of the original photon. Consequently the original absorption wavelength is missing, or nearly so, in the light that arrives on Earth and a Fraunhofer line appears in the spectrum. The lines are very sharp because if the photon frequency is not extremely close to that corresponding to the transition energy it will not be absorbed — it is a resonance effect. The two very close prominent orange-yellow lines seen in the sodium spectrum in Figure 6.7, known as the sodium D-lines with wavelengths 588.9960 nm and 589.5924 nm, correspond to the close pair of lines marked D in Figure 6.6.

6.4. The Composition of Stars

The Fraunhofer lines seen in the spectrum of the Sun also appear in the light from other stars. We have already commented that if the sodium *D*-lines are seen, either as bright emission lines or as absorption lines, then we know that they are due to the presence of sodium and nothing else. From the depth of the absorption lines (Figure 6.8), in essence how dark they are, we can deduce the projected areal density of sodium in the path taken through the absorbing region. Projected areal density is the total mass per unit area integrated through the absorbing region, i.e. the total amount of sodium contained in the volume shown in Figure 6.9.

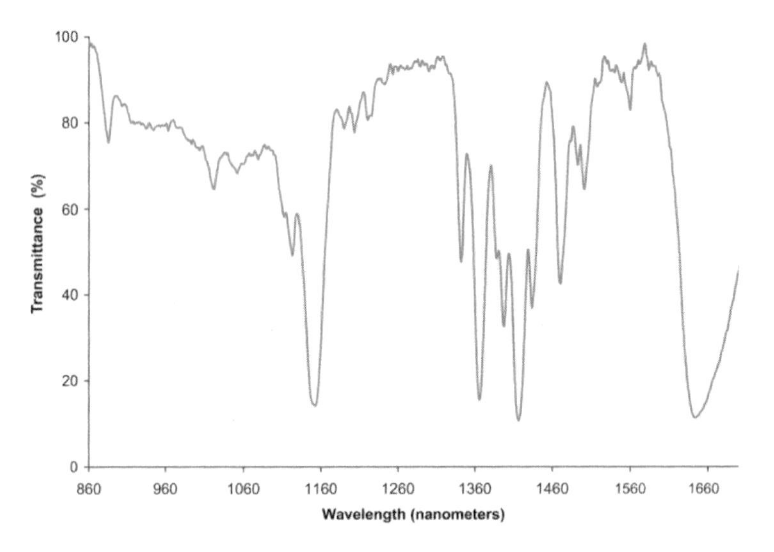

Figure 6.8. A stellar spectrum showing absorption lines.

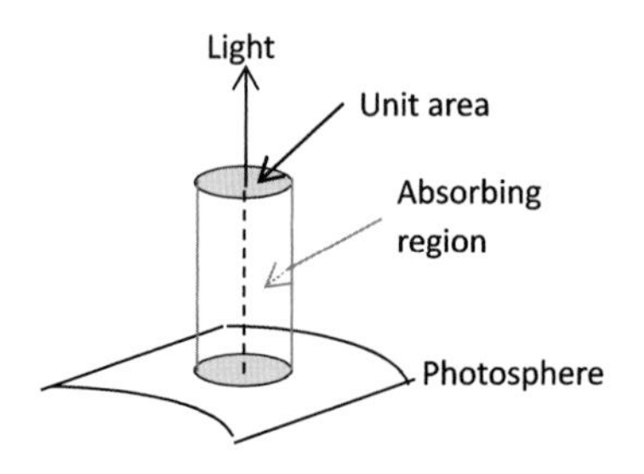

Figure 6.9. The projected areal density.

Every type of atom and ion has its characteristic absorption spectrum and by measuring the intensity of the lines it is possible to estimate the amounts of the various elements in the absorbing region, which is taken to represent the composition of the star.

At very high temperatures molecules are broken down, either into individual atoms or small stable molecular fragments. Even at the solar surface temperature, just under 6,000 K, some stable molecular fragments exist and in low temperature stars there are considerable quantities of molecular material. As mentioned in Section 6.1 there are discrete energy levels for molecules, corresponding to different conformations they can adopt, and absorption lines corresponding to transitions between these states give Fraunhofer lines in stellar spectra.

6.5. Metallicity

Stars do not all have the same composition although all main-sequence stars are predominantly hydrogen and helium. There are various estimates for the composition of the Sun, as judged by what is observed in the visible outer layers, but is probably, by fractional mass, hydrogen 0.74, helium 0.24 and other elements 0.02. For stars in general, astronomers will refer to these fractions as X, Y and Z respectively, and Z is called the *metallicity* of the star's content even though that fraction includes elements such as carbon, oxygen and sulphur that are not metals.

The pattern that is suggested by our description of the evolution of the Universe and the stars within it, is that stars in globular clusters should have a metallicity of zero, because they consist of primordial material from the Big Bang, mainly hydrogen and helium with traces of the light elements lithium and, perhaps, beryllium. When the most massive of these stars — the ones that evolve most quickly — go to the supernova stage they inject heavier elements into the interstellar medium of the galaxy of which they are a part and the galactic-cluster stars, formed later, therefore have larger metallicity. The younger a star is, the larger should be its metallicity because more and more stars, both in globular and galactic clusters will have progressed to the supernova stage prior to its formation. Some material in young stars may have been processed by several generations of former stars with the metallicity of the material increasing with every generation.

The conclusion that the younger is a star the higher will tend to be its metallicity is true but what is not true is that globular-cluster stars have zero metallicity. Stars are divided into two categories Population I, which are the high-metallicity stars found in galactic clusters, and Population II, which are the low-metallicity stars found in globular clusters. Because of the finite metallicity of globular-cluster stars, astronomers have postulated that there were Population III stars that were formed from primordial material and that these provided the metal content of the Population II stars. They have searched for, but failed to find, any such stars. It seems more likely that some early massive stars, which quickly evolved, were present in the evolution of globular clusters and the stars formed later derived their metal content from that source. The metallicities of Population II stars usually fall in the range from one-thousandth to one-hundredth that of the Sun.

If the stellar spectrum is not of good quality it may be difficult to estimate the amount of some elements. In that case an alternative way of defining metallicity can be used. Even in poor spectra the iron lines are usually quite clear so it is possible to measure the ratio of the amount of iron to hydrogen, the abundance of which is also comparatively easy to measure. The measure of metallicity used is then

$$[\mathrm{Fe/H}] = \log\left(\frac{(\mathrm{Fe/H})_\circ}{(\mathrm{Fe/H})_\odot}\right). \tag{6.3}$$

This compares the ratio of iron to hydrogen in the star to that in the Sun, if [Fe/H] is zero then the metallicity of the star equals that of the Sun. If the value is -1 then it is one-tenth that of the Sun and if 1 it is ten times that of the Sun. The latter value would be a very metal-rich star, presumably a very recently-produced Population I star. By contrast a value of -4 would correspond to a very metal-poor Population II star.

Problems 6

6.1 Two energy levels for mercury are $-10.38\,\text{eV}$ and $-5.74\,\text{eV}$. Find the wavelength of the Fraunhofer line corresponding to a transition between these two states. In what part of the electromagnetic spectrum does it occur?
$(1\,\text{eV} = 1.602 \times 10^{-19}\,\text{J})$

6.2 The proportions of hydrogen, helium and iron by mass in the Sun are 0.7381, 0.2485 and 0.0019. What is the metallicity of the Sun?

A star has proportions by mass of hydrogen, helium and iron 0.726, 0.252 and 0.00306. Find two estimates for the metallicity of the star.

Part 4

The Distances of Stars

Chapter 7

Finding the Distances of Nearby Stationary Stars

When determining quantities associated with an object, such as its mass or size it is convenient, if possible, to have units that express the quantities as small numbers. A 6 tonne elephant is easy to understand; objects like motor cars weigh about one tonne and the weight of six cars is easily envisaged. However, if the weight of the elephant was given as 6 million grams, equivalent to 6 tonnes, then the only comparison would be something like the weight of a sheet of standard A4 paper, about 5 grams and it is impossible to envisage the weight of an elephant as equivalent to that of 1,200,000 sheets of paper. As another example we take the case of the man who, unassisted, travelled 100,000 millimetres in 1.110×10^{-4} days. Is he an athlete or someone who is disabled? He is in fact the Jamaican athlete Usain Bolt and what has been described is his 2009 world record for the 100 metres, 9.59 s.

In Section 3.1, the *parsec*, equivalent to 3.26 ly was mentioned as a convenient unit for measuring large distances. In this chapter, we shall see how this unit arises naturally from the method used to measure the distances of the nearest stars.

7.1. How Far Away is That Church Steeple?

Consider a situation where it is required to measure the distance from some fixed point to the pinnacle of the steeple of a church situated in the valley below. Between the fixed point and the church there is rough ground, a river and several substantial hedges so the idea of using a surveyor's steel tape to measure along the ground is not practical. In any case, when the base of the church was reached it would not be possible to measure the horizontal distance to the steeple. Some other method is required.

In addition to the steel tape a theodolite is available, an early form of which is shown in Figure 7.1. It is basically a telescope, fitted with fine cross hairs, which can be rotated about both a vertical axis and a horizontal axis. The angular position of the telescope with respect to rotation about the vertical axis can be read on a horizontal circle to an accuracy of one-tenth of a second of arc. To put this in perspective it is approximately the angle subtended by 1 cm at a distance of 20 km! There is also provision to measure position with respect to rotation about the horizontal axis but we shall assume that the steeple pinnacle is at the same level as the fixed point so that only horizontal angular measurements are involved.

Now the steel tape is used to establish a new point, B, at some distance, d, of order 50 m from the fixed point, A, such that AB is approximately perpendicular to AS, where S is the position of the pinnacle (Figure 7.2). The theodolite is mounted on a tripod and a plumb bob ensures that the vertical axis is directly over the point A. The telescope is rotated until the cross hairs intersect a narrow vertical rod fixed at B and the reading is taken on the horizontal circle. The telescope is then rotated until the cross hairs intersect the pinnacle of the steeple and a second reading is taken. The difference of the two readings gives the angle α in Figure 7.2 so we know that the steeple is somewhere on the line AP. Moving the theodolite to B we can, in a similar way, find angle β and this then locates S at the intersection of AP and BQ. This process is known as *triangulation* and if the length of the *base line*, d, and the angles α and β are known then it is clear that the distance AS can be determined.

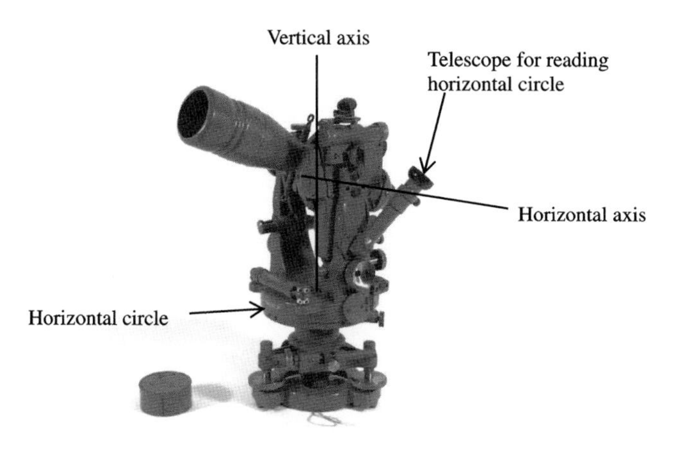

Figure 7.1. The Cooke, Troughton and Simms Tavistock theodolite.

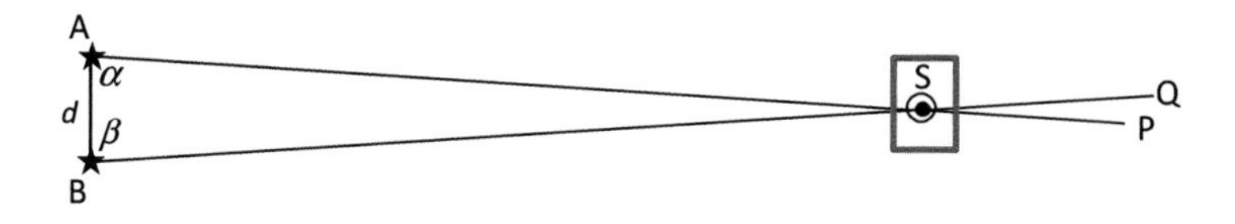

Figure 7.2. Triangulation from the base line AB to find the position of S.

Triangulation plays an important role in cartography — map-making. Using a very precisely measured baseline AB the position of point C is determined, where ABC is as close as possible to an equilateral triangle. Then AC and BC can be used as base lines to find the positions of new points. With a suitably long and precise initial base line, of length some tens of kilometres, a large area, even a whole country, can be covered by what is known as a system of *primary triangulation*. With these points accurately fixed within each primary triangle there is secondary triangulation and within each secondary triangle there is tertiary triangulation. The detailed topography is then determined within the tertiary triangles. Figure 7.3 shows the primary triangulation of Ireland, completed in 1969. Until the last years of the twentieth century mapmakers had only steel tapes and theodolites to measure baselines and to find the positions of landscape features. Later lasers were used to measure baselines by

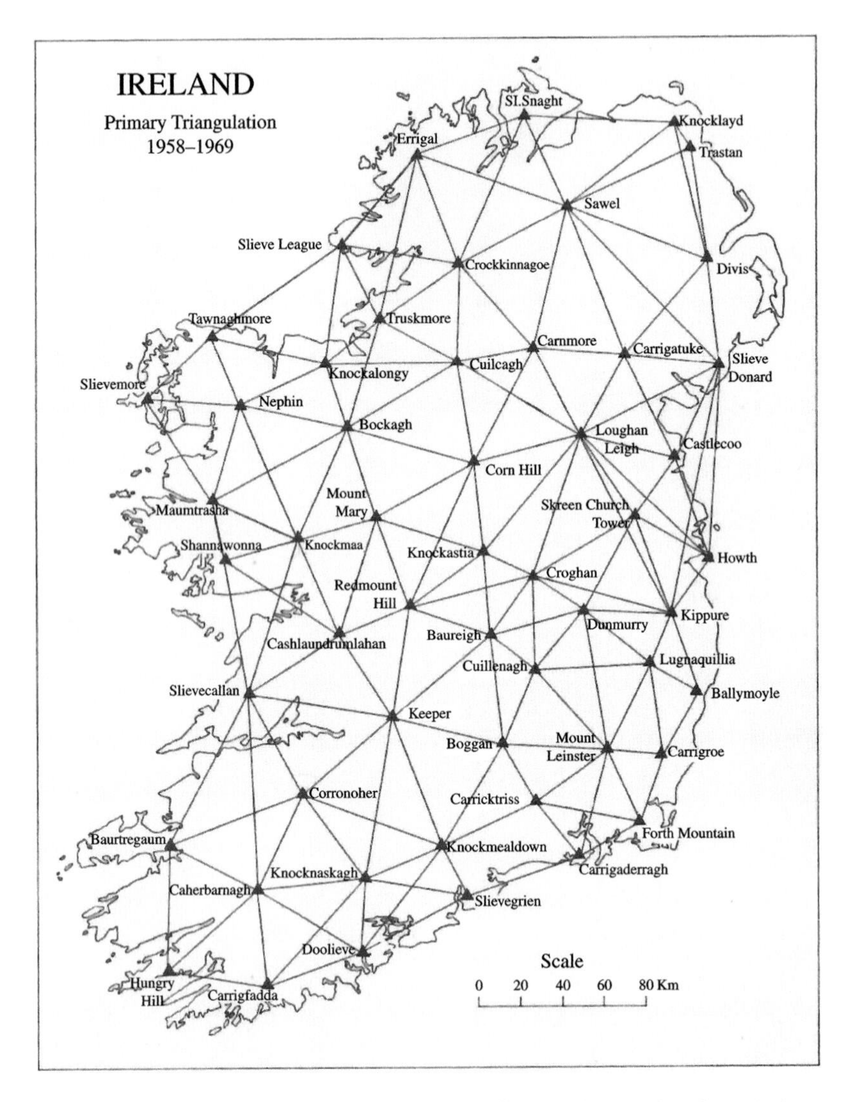

Figure 7.3. The primary triangulation of Ireland, completed in 1959.

measuring the time for a very short light pulse to travel from and to one end of the baseline having been reflected from the other end. This was a very precise method; reflecting light from retro-mirrors left on the Moon by Apollo astronauts enables the very slow rate of recession of the Moon from the Earth to be measured. Nowadays

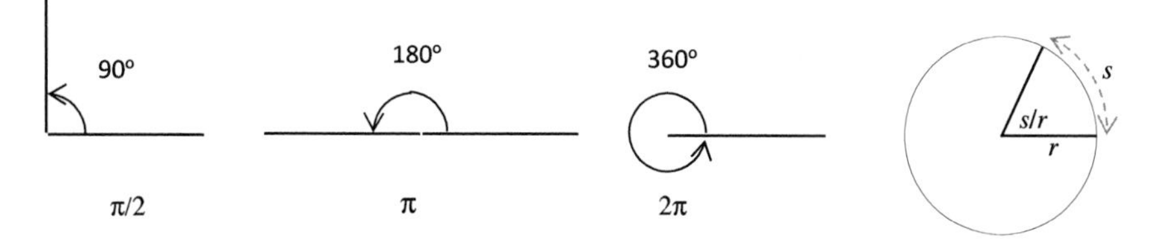

Figure 7.4. Angles expressed in degrees and radians.

surveying relies heavily on GPS (Global Positioning System) that uses signals from several satellites to locate positions on Earth.

7.2. Radians and Small Angles

The usual school introduction to the concept of an angle is that it is measured in degrees. Ninety degrees (90°) is a right-angle, 180° looks like a straight line and 360° is a complete rotation (Figure 7.4). However, for many scientific applications there is another, more convenient way of expressing angles, also illustrated in Figure 7.4, where the angle shown is described as s/r *radians*, where r is the radius of the circle and s the length of the arc that embraces the angle. Since the circumference of a circle is $2\pi r$ then 360° is 2π radians, 180° is π radians and 90° is $\pi/2$ radians. One radian, when $s = r$ is equivalent to 57.29°.

The process of determining the distance of nearby stars involves the determination of very small angles, of the order of seconds, or fractions of seconds, of arc.[1] The kind of triangle we shall be considering is illustrated in Figure 7.5, although the angle α shown is much larger than it would be in an actual measurement. The length of line AB (s) is virtually the same as the length of an arc of radius OA (r) linking A and B so, from the definition of an angle expressed in radians

$$\alpha = s/r. \tag{7.1}$$

[1]A degree is divided into 60 min of arc (symbol′) and each minute of arc into 60 s of arc (symbol″)

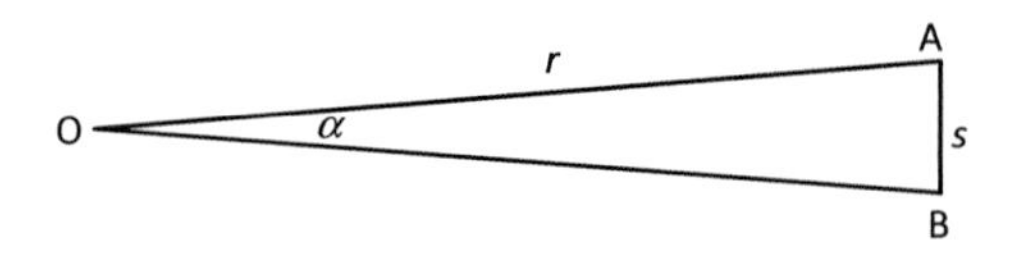

Figure 7.5. The length s closely equals the length of the arc linking A and B centred on O if α is very small.

7.3. How Far Away is That Stationary Star?

We have seen that with a well-measured base line and a theodolite we can determine the position of a distant object with respect to any point on the base line, and that this can be done without physical access to the intervening space. Now we shall see how to apply this general principle — observing from two locations — to determine the distance of a nearby star. To do this a phenomenon known as *parallax* is used. To illustrate this, face a distant scene and hold a finger vertical at arm's length. If you close your left eye then with your right eye you see your finger superimposed on a feature of the distant scene. Now close your right eye and then with your left eye you see your finger superimposed on a different feature of the distant scene. Alternately closing first your left eye and then your right, your finger seems to jump between the two points on the distant scene. Next bring your finger closer to your face and repeat the experiment. What you find is that the two points on the distant scene are further apart. It seems that from the distance between the two points on the distant scene and, another obvious factor, the separation of your eyes it should be possible to find the distance of your finger from your face.

The idea of using parallax to measure the distance of nearby stars was first proposed by the brilliant German mathematician and astronomer, Johannes Kepler (1571–1630; Figure 7.6). For the two viewing points he proposed the positions of the Earth in its orbit six months apart, i.e. at opposite ends of a diameter of the Earth's orbit, and for the distant scene the background of very distant stars, many thousands of times further away than the star whose distance is to be determined. Kepler tried to use his method but the technology of

Figure 7.6. Johannes Kepler.

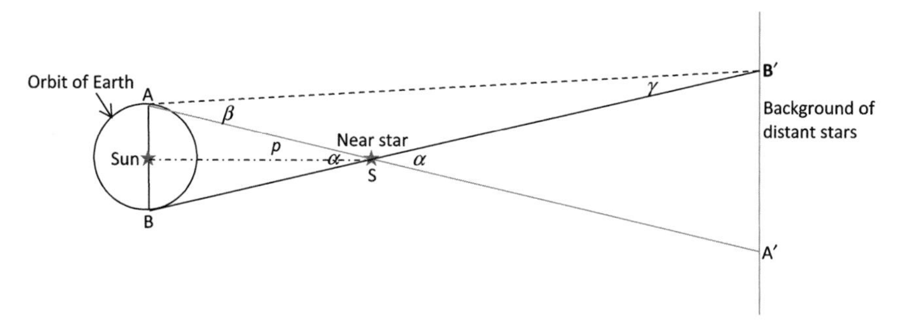

Figure 7.7. The geometry of the parallax method.

the time was not good enough for him to succeed, although now it can be used successfully.

The geometry of the *parallax method* is illustrated in Figure 7.7, where it is assumed that the star does not move between the observations. The observing points on the Earth's orbit, A and B, are chosen so that AB is perpendicular to the direction of the star. The positions of the star, as seen on the background of very distant stars, are A′ and B′.

Before we go any further it is useful to look at some orders of magnitude of the lengths and angles displayed in Figure 7.7. The mean Earth–Sun distance, the astronomical unit, is, 1.496×10^8 km and the distance AB is 2 au. The angle α is of the order one second of arc. The background stars are hundreds of thousands or more light years from the Sun. For the triangle AB'S the exterior angle α equals the sum of the two interior angles β and γ. However, the angle γ is extremely small. Using equation (7.1) and assuming the far stars are at distance 10,000 ly, it is

$$\gamma = \frac{2 \times 1.496 \times 10^8}{100,000 \times 9.461 \times 10^{12}} = 3.162 \times 10^{-10} \text{ radians.}$$

Multiplying this by 57.29 gives the angle in degrees and then multiplying by 3,600 gives it in seconds of arc. This gives $\gamma = 6.521 \times 10^{-5''}$. An observation is recorded photographically and the parallax angle is then determined from the motion of the near star between positions A and B. The accuracy with which parallax angles can be measured in this way by ground-based telescopes, under the best conditions, is of order $5 \times 10^{-3''}$ but more usually $0.01''$. The limitation for ground-based measurement is atmospheric fluctuations which disturb the path of a light beam. It can be seen that γ is very much smaller than α and for that reason we can take β, which comes directly from the photographs, as a good approximation for α.

The parallax method has given rise to a unit of measurement for stellar distances that is usually used, rather than light-years. The angle $p(= \alpha/2)$ shown in Figure 7.7 is known as the *parallax angle.* If this angle is $1''$ then the distance is 1 *parsec* (symbol pc); it is the distance at which 1 au subtends an angle of $1''$. The relationship between the distance of a star, D, and its parallax angle is

$$D = 1/p = 2/\alpha \approx 2/\beta, \tag{7.2}$$

and D is in parsecs if p is in seconds of arc. The parsec equals 3.26 ly, or 3.0857×10^{13} km.

The nearest star, Centauri Proxima, is at a distance of 1.301 pc (4.25 ly) so the parallax angle is 1/1.301 or $0.769''$. If the possible error is, say, $\pm 0.005''$ then the estimate of distance has a possible error of about 0.01 pc. The maximum stellar distance measured from ground observations is about 100 pc, for which $p = 0.01''$ and with the same accuracy of measurement the estimate of distance could be in error by as much as 50%.

For reasonable accuracy, say ±20%, ground-based measurements are capable of giving the distances of a few hundred stars out to about 50 pc.

7.4. Space-based Measurements

In 1989, the European Space Agency (ESA) launched a space mission, Hipparcos, designed to measure many properties of stars, including their distances. The name Hipparcos is an acronym for "HIgh Precision PARallax COllecting Satellite" but sounds similar to Hipparcus, a second century BCE Greek astronomer and mathematician who is credited with the invention of trigonometry, which is the basis of the parallax method, and also with producing a star map containing over 1,000 stars. Freed from the limitations of ground-based observations due to the atmosphere, parallax angles can be estimated with an accuracy of $0.001''$ and this greatly extended the range of measurement while preserving reasonable accuracy. Hipparcos ceased operating in 1993 and in 1997, after a considerable amount of data processing, the Hipparcos Catalogue was published giving high-precision distances for 118,218 stars (Turon, C. *et al.*, (1995). The error is less than 10% for 20,000 stars within a distance of 100 pc and less than 20% for those within a distance of 200 pc, accounting for about 70,000 stars.

In 2013, ESA launched the space mission Gaia, a higher-precision successor to Hipparcos. With a precision of about $2 \times 10^{-5''}$ in the measurement of parallax angles — corresponding to the width of a human hair seen at a distance of 1,000 km — it should give distances for about 20 million stars within a distance of 10,000 pc to an accuracy of 20% or better.

Problem 7

7.1 A Gaia measurement for a stationary star gave the angle β (see Figure 7.7) as $3.72 \pm 0.02 \times 10^{-3}$. Find the distance of the star in pc and the expected precision of the measurement as a percentage error.

Chapter 8

Finding the Distances and Velocities of Nearby Moving Stars

8.1. Speed and Velocity

In everyday language the terms 'speed' and 'velocity' tend to be used interchangeably as though they were synonyms. However, in a scientific context they are different; speed implies no particular direction and if we say that a car is travelling with a speed of $60\,\text{km h}^{-1}$ then it could be going in any direction. By contrast velocity does imply direction and we might say that a car has a velocity of $60\,\text{km h}^{-1}$ northeast. Another car moving at $60\,\text{km h}^{-1}$ south has the same speed but a different velocity. In mathematical language speed, which is non-directional, is a *scalar quantity* and velocity, which has an associated direction, is a *vector quantity*. We can represent a vector quantity by an arrowed line, the length of which indicates its magnitude with direction given by that of the line (Figure 8.1). A vector can have *component vectors*; for example, if a car is moving northeast we can have a northward component vector and an eastward component vector, as shown in Figure 8.1. However, the number of possible pairs of components is infinite and an arbitrary pair of components is shown in the figure. For our present purpose it should be noted that when we refer to the velocity of a star we mean both its speed and direction of motion.

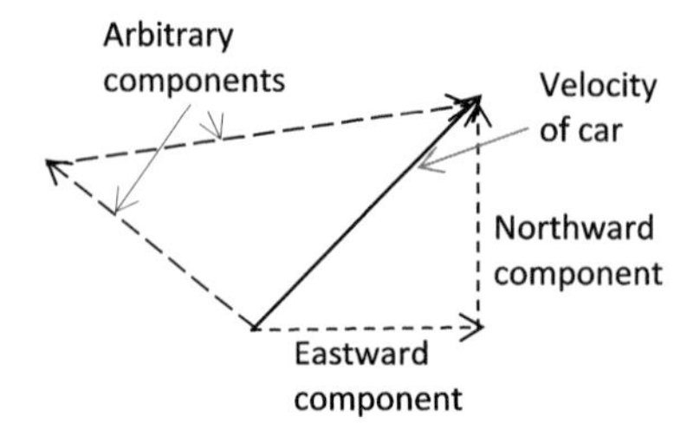

Figure 8.1. A vector and two possible pairs of components.

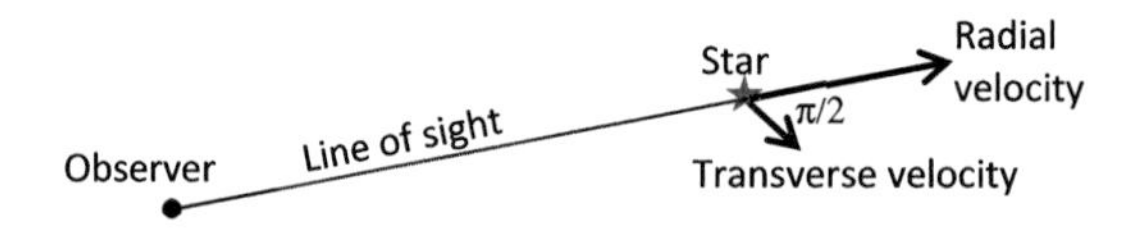

Figure 8.2. The radial and transverse velocity components of a star's motion.

8.2. The Components of a Star's Velocity

So far we have just considered the problem of finding the distance of a nearby stationary star but, actually, the stars in our galaxy are all moving relative to each other, and to the Sun, with average relative speeds of 20–30 km s^{-1} in all possible directions. We shall see that it is possible to determine by observations two components of the velocity of a star, one, the radial velocity, along the line of sight, i.e. directly away from or towards the observer on Earth, and the other, the transverse velocity at right angles to the radial velocity (Figure 8.2).

8.3. Finding the Distance and Transverse Velocity of a Nearby Star

To find both the distance and transverse velocity of a nearby star it is necessary to make three observations, at A, at B and at A again, when the positions of the star are at S_1, S_2 and S_3, respectively (Figure 8.3).

It is important when viewing this figure to appreciate that, for clarity, angles are shown much larger than they really are and

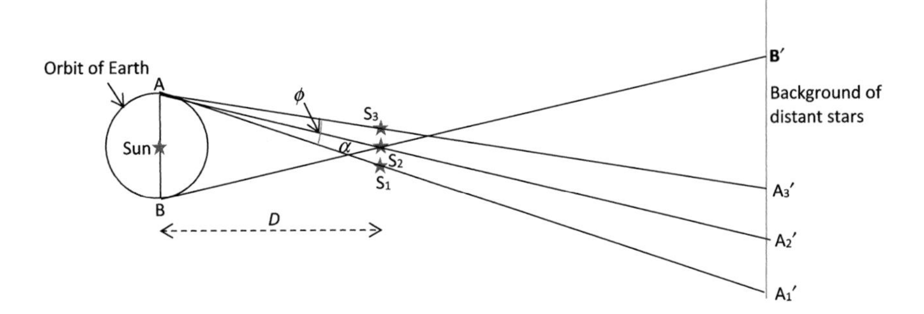

Figure 8.3. The geometry of parallax with a moving star.

horizontal distances are truncated. Thus, in reality, all the lines running from left to right are nearly parallel to within a second of arc or less. The positions of the nearby star, seen against the background of distant stars, A_1' and A_3' are found one year apart when the star is at S_1 and S_3, respectively. The position of the star when the observation from B is being made is at the midpoint of S_1S_3, marked as S_2. When the star was at that position it would have been seen from A at A_2', the midpoint of $A_1'A_3'$, which can be found because the positions A_1' and A_3' were recorded. Now A_2' and B' play the role of A' and B' in Figure 7.7 and so the angle α and hence the distance of the star, D, can be found.

Because the star has a radial velocity its distance from the observer is constantly changing so we should check that the change of distance between positions S_1 and S_3 is negligible compared to D. If the radial speed is $100\,\text{km s}^{-1}$, a very high value, then in one year, $3.156 \times 10^7\,\text{s}$, the distance will have increased or decreased by 0.0001 pc, less than the expected error in most measurements.

With the known locations A_1' and A_3', both recoded from A, on the background of distant stars the angle ϕ can be found. The distance moved by the star in the transverse direction in one year is then given by $S_1S_3 = D\phi$ from which the transvers speed is $D\phi/T_y$, where T_y is 1 year. The direction of motion of the star in the transverse direction is indicated by the line S_1S_3 and with both speed and direction known the transverse velocity has been defined.

8.4. Determining the Radial Velocity of a Star

The laws of physics apply everywhere in the Universe so Fraunhofer lines that occur in the solar spectrum (Section 6.3) will also occur in the light from other stars. While estimating the radial speed by finding the wavelength shift of the whole spectrum would be difficult and imprecise, nature has provided us with the very well-defined Fraunhofer lines, the Doppler wavelength shifts of which can be measured with high precision by spectrometry. From equation (2.40), with V equal to c, the speed of light, the radial velocity comes from

$$v = \frac{cd\lambda}{\lambda}. \tag{8.1}$$

If the sodium D-line, with wavelength 588.9950 nm measured in the laboratory, is measured from a star with wavelength 589.0282 nm then, using equation (8.1)

$$v = 2.998 \times 10^5 \times \frac{589.0282 - 588.9950}{588.9950} = 16.9\,\text{km s}^{-1}.$$

Since the measured wavelength from the star is greater than the laboratory value the star is moving away from the Earth (see Figure 2.11).

The Earth's average speed in its orbit around the Sun is 29.78 km s^{-1} so the measured radial velocity of a star would depend on where the Earth was in its orbit. What is required is the radial velocity of the star relative to the Sun. To do this we use the following relationship Radial velocity of star relative to the Sun = Radial velocity of star relative to the Earth + component of Earth velocity relative to the Sun in the direction of the star.

If in Figure 8.3 the line AB is perpendicular to the direction of the star then at both measurement points the Earth has no component of velocity in the direction of the lateral motion of the star. The lateral speed with respect to the Sun is then the same as that relative to the Earth. However, it has orbital speed along the line of sight.

With both the lateral and radial velocity components known the overall velocity of the star, i.e. its speed and direction, are known relative to the Sun.

Problem 8

8.1

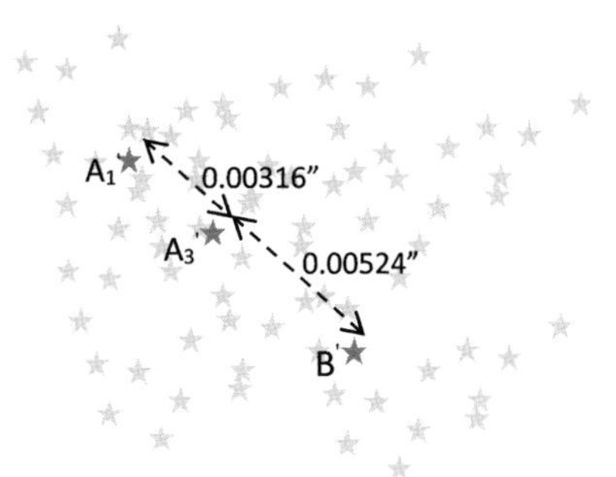

The figure shows the three positions of a nearby star against the background of distant stars with positions labelled as in Figure 8.3. The angular difference between pairs of points, in seconds of arc, are indicated. What is the distance of the star and its transverse velocity?

A sodium line has a wavelength of 588.9950 nm. At position A_1, when the Earth is moving away from the star, the wavelength measured is 588.9791 nm and at B the measured wavelength is 589.0913 nm. Estimate the radial speed of the star with respect to the Sun. Take the orbital speed of the Earth as its mean value, 29.78 km s^{-1}.

Chapter 9

Finding Distances to Faraway Stars

9.1. Finding the Temperatures of Stars

The distribution of wavelengths radiated by stars is temperature dependent, as is seen in Figure 2.9. This means that, in principle, by just considering the variation of intensity with wavelength of the radiation coming from a star, we could estimate its temperature. In practice this would not be very precise. For one thing the curves shown in Figure 2.9 are theoretical curves for what is known as a *black body*, an ideal body that absorbs all the radiation falling on it. While stars approach that ideal they are not perfect black bodies. Another problem is that dust between the star and Earth scatters light, blue much more than red. The scattered light does not reach the Earth so light from a star will appear redder than it was when it was emitted and the extent of the reddening is distance dependent Despite this difficulty, a reasonably good estimate of temperature can be made by comparing the relative intensity of the light passing through a standard pair of filters, illustrated by the red and blue filters in Figure 9.1.

Astronomers have found a much more sensitive method of estimating temperature, based on the energy levels in atoms, ions and molecules. As an example we consider hydrogen. The ground state, i.e. the lowest energy level, of a hydrogen atom, is $-13.6\,\text{eV}$, which means that if $13.6\,\text{eV}$ or more of energy were given to the electron then its energy would become positive and it would escape

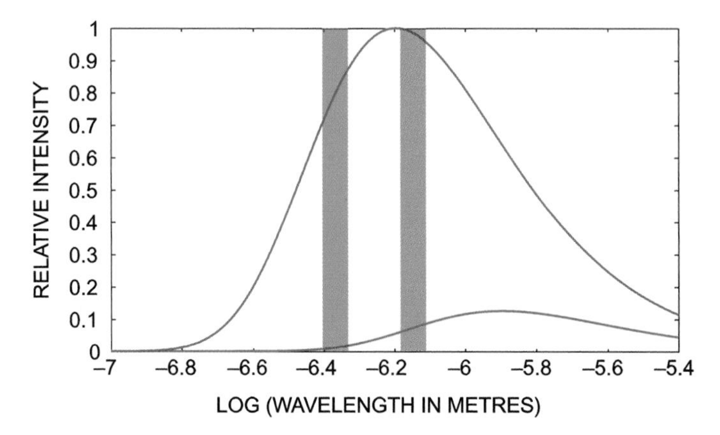

Figure 9.1. The passage of light through a red and blue filter for temperatures 4,000 K (red) and 8,000 K (blue).

from the atom to become a free electron. Successive energy levels are:

$$E_1 = -13.6\,\text{eV} \quad E_2 = -3.4\,\text{eV}$$
$$E_3 = -1.5\,\text{eV} \quad E_4 = -0.85\,\text{eV}, \ldots, E_n = -13.6/n^2\,\text{eV}.$$

The energy levels get ever closer with increasing n and eventually merge with the ionization level, when the electron just escapes.

For radiation coming from a body at temperature T the average photon energy increases with T and is of order kT, where k is the *Boltzmann constant* (1.381×10^{-23} J s^{-1} = 8.620×10^{-5} eV s^{-1}). However, there is a distribution of photon energies about the mean and there will be many photons with more than three times the average energy. Let us now consider the transition for hydrogen from E_1 to E_2, which requires the electron to receive energy from a photon with energy 10.2 eV. For a low-temperature star there will be very few photons with sufficient energy to give the transition and so there will be no dark line in the spectrum corresponding to that transition. Increasing the temperature will give more and more photons with sufficient energy and so the line will become stronger. At some temperature the increased probability of the transition from E_1 to E_2 for every atom with an electron in the ground state will be balanced

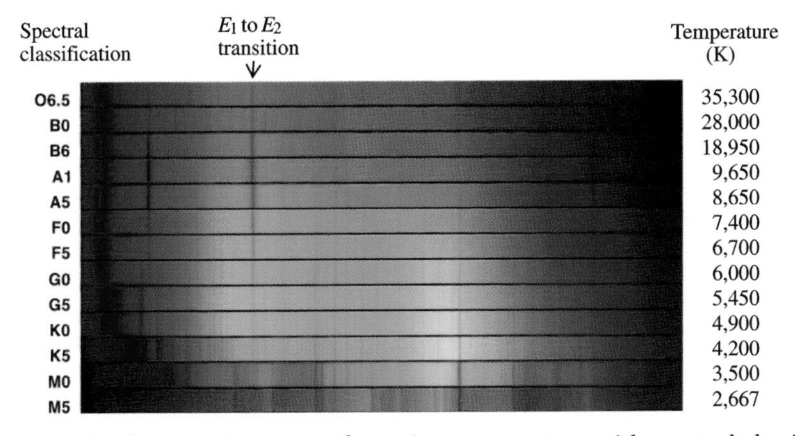

Figure 9.2. A spectral sequence for main-sequence stars with spectral classification and temperatures (H bond (STScI)).

by the fact that so many hydrogen atoms have become excited away from the ground state that there are fewer ground state electrons available. Thenceforth as the temperature increases the absorption line becomes weaker and at a very high temperature will disappear. Other kinds of absorption line for hydrogen and other atoms and ions have similar properties — appearing, strengthening, weakening and disappearing as the temperature increases for a particular transition. Figure 9.2 shows a sequence of spectra at different temperatures; with many more temperatures included it would be seen that there is a smooth change in the appearance of an absorption spectrum with temperature. Theoreticians have calculated the way that different absorption lines should vary in intensity with temperature. The appearance of the complete set of absorption lines — some missing, some weak and some strong — gives a good indication of the temperature and this is the method of choice for determining the temperature of a main-sequence star.

From the appearance of the spectra astronomers have devised a *spectral classification* of stars with letters O, B, A, F, G, K, M (mnemonic: *Oh Be A Fine Girl, Kiss Me*) indicating stars from O, the hottest, to M, the coolest. Each class is divided into ten subclasses by numbers 0 to 9 so class G gives $\mathrm{G0}, \mathrm{G1}, \ldots, \mathrm{G9}$, going from the hottest to the coolest. The Sun is a G2 star with temperature 5,772 K.

There are types of stars other than main sequence and at a particular temperature they give similar absorption lines, but different widths of those lines due to different densities and compositions of the atmospheres giving rise to the absorption. The type of star is indicated by adding an appendage to the spectral class. These are:

0 for hypergiants, I for supergiants, II for bright giants, III for normal giants, IV for subgiants, V for main-sequence stars (dwarfs), and for sub-dwarfs and *D* for white dwarfs.

Thus a full description of the spectral classification of the Sun is G2V and of Canopus, a bright giant, A9II.

9.2. Luminosity, Magnitude and Brightness

Sometime in the second century BCE the Greek astronomer Hipparchus produced a catalogue in which stars were numerically assigned *magnitudes* running from 1 to 6 with 1 corresponding to the brightest stars. By the 18th century astronomers had refined their judgement of magnitudes so that instead of having only integer values one could refer to the magnitude of a star as 4.4, for example. In the 19th century instruments became available to give quantitative measurements of the brightness of stars as energy per unit area per unit time received on Earth. It was found that the Hipparchus range from 1 to 6, i.e. five units of increment, corresponded to a factor of about 100 in brightness and a new scientifically-based scale of magnitude (symbol m) was established in which a difference of 5 in magnitude corresponded precisely to a change in brightness by a factor of 100. In this scale each increment of 1 gave the same factor of decrease, which was $10^{2/5}(= 2.512)$, so that a change of magnitude of 5, from 1 to 6, gave a decrease of brightness of

$$\frac{b_1}{b_6} = \frac{b_1}{b_2}\frac{b_2}{b_3}\frac{b_3}{b_4}\frac{b_4}{b_5}\frac{b_5}{b_6} = \left(10^{2/5}\right)^5 = 10^2 = 100, \tag{9.1}$$

where b_m is the brightness of a star with magnitude m.

With modern telescopes and with very sensitive CCD (charge-coupled device) detectors the range of measured magnitudes has been greatly extended. Very bright objects have negative magnitudes so that the brightest object in the sky, the Sun, has magnitude -26.74 and the bright star Sirius has magnitude -1.46. With a large telescope and long observing times the largest observable magnitude (i.e. corresponding to the faintest star) is somewhere in the range 26 to 28. To fix a scale of magnitudes the bright star Vega is taken to have a magnitude of 0. A general expression for comparing the brightness of two stars of magnitude m_1 and m_2 is

$$\frac{b_1}{b_2} = (10^{2/5})^{m_2 - m_1} = 10^{2(m_2 - m_1)/5}. \tag{9.2}$$

Thus, a star with magnitude 26 would be fainter than Vega by a factor $(10^{2/5})^{26} = 2.512 \times 10^{10}$.

Now we consider the relationship between the luminosity (Section 4.1) and brightness. Let us first consider it in terms of car headlights. They will have a luminosity given by their total power output. The undipped headlights of an approaching car at a great distance will not seem very bright but as it approaches the brightness increases until, close up, it would be dazzlingly bright unless the driver dipped the lights. Clearly, in any relationship between brightness and luminosity, distance must play a role.

A star of luminosity L_*, unlike headlights, radiates its energy uniformly in all directions. At distance d an amount of energy per unit time L_* crosses an area $4\pi d^2$, the surface area of a sphere of radius d. Thus the *apparent brightness* — i.e. energy received per unit area per unit time — at distance d is

$$b_d = \frac{L_*}{4\pi d^2}. \tag{9.3}$$

A measure of brightness that, like luminosity, is an inherent property of a star that enables comparisons to be made between stars, is its *intrinsic brightness* (symbol B), its brightness if it were seen at a standard distance, which is taken as 10 pc. Hence we may write

$$B = \frac{L_*}{4\pi \times 10^2}, \tag{9.4}$$

that, with (9.2), gives

$$\frac{b_d}{B} = \frac{100}{d^2}, \tag{9.5}$$

where d is in parsecs.

Another observed property of a star that depends on distance is what we have called magnitude but which we will now call *apparent magnitude* since is depends on the distance from which the star is viewed. Corresponding to intrinsic brightness, sometimes called *absolute brightness*, we define *absolute magnitude* (symbol M), the magnitude of a star when seen at a distance of 10 pc. Now using equations (9.2) and (9.5)

$$\frac{b_d}{B} = \frac{100}{d^2} = 10^{2(M-m)/5}. \tag{9.6}$$

Taking logarithms of both sides of the final equation of (9.6)

$$2 - 2\log d = 2(M - m)/5.$$

Rearranged, this gives

$$M = m + 5(1 - \log d), \tag{9.7}$$

that turns out to be a very useful relationship for finding the distances of stars.

9.3. Distance Measurement Out to 10,000 pc Using Main-Sequence Stars

Table 4.1 shows that the main-sequence is a very long-lasting state of a star and, for that reason, the majority of stars that are observed are in that state. During the main-sequence lifetime the properties of the star will change very little. From observations of stars within the parallax range it is found that all main-sequence stars with the same temperature, a property we can measure regardless of distance, have the same absolute magnitude. The absolute magnitudes are derived from equation (9.7) because the parallax method gives d and brightness measurement gives m.

With absolute magnitudes known from the spectral class, measurement of apparent magnitude gives the distance. As an example we take the star Saiph. The absolute magnitude is −6.98 and the observed apparent magnitude is 2.07. From equation (9.7)

$$\log d = (m - M)/5 + 1 = (2.07 + 6.98)/5 + 1 = 2.81,$$

giving

$$d = 10^{2.81}\,\text{pc} = 645\,\text{pc}.$$

This method of finding distances can be used out to 10,000 pc or so but with increasing proportional error with distance as it becomes more difficult accurately to assess the spectral features of stars of large apparent magnitude. However, when there is a galactic cluster containing a large number of main-sequence stars the aggregate information from them can give a more precise distance estimate. One way of representing stars on an H–R diagram is shown in Figure 9.3 where the x-coordinate is log(temperature), as in Figure 4.1, but the y-coordinate is magnitude. The line marked 'main sequence' gives the absolute magnitudes of stars on the main sequence; for illustration it is given as a straight line although it is actually curved. The various

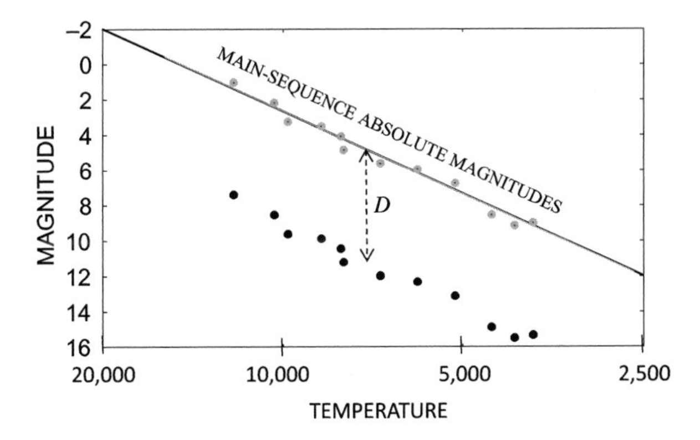

Figure 9.3. A schematic representation of using cluster apparent magnitudes to find a cluster distance.

black dots give the apparent magnitudes of a number of stars in the cluster; in practice there would be many more since galactic clusters can contain many hundreds of stars. The grey dots represent a displacement of the black dots to fit the main sequence line as well as possible. The distance marked D is a best average of $m - M$ that, from equation (9.7), is $5(\log d - 1)$. The distance found will be an estimate of the distance of the centre of the cluster but, generally, the sizes of clusters are small compared to their distances.

For this particular example $D = 6.0$ and hence $\log d = 2.2$ or $d = 158\,\text{pc}$.

9.4. Distance Measurement Using Cepheid Variables

The groundwork for the next method of measuring stellar distances was established in the 18th century due to observations made by the English astronomer John Goodricke (1764–1786; Figure 9.4(a)) on variable stars — stars that vary in intensity in a periodic way. Despite the social handicap in those days of being a deaf mute, Goodricke became educated, took up astronomy and discovered a number of important variable stars. The first of these, Algol, observed in 1782, was an eclipsing binary in which two stars, too close together to be individually resolved and of different intrinsic brightness but similar size, eclipsed each other at regular intervals.

(a) (b)

Figure 9.4. The Cepeid-variable pair (a) John Goodricke (b) Henrietta Leavitt.

With respect to finding distances the most significant discovery by Goodricke was his last one, the variable star δ-Cephei which fluctuates in brightness with a period of about 5.4 days (Goodricke, 1786). It is said that his early death, at the age of 21, was caused by pneumonia contracted while making observations of δ-Cephei. This star is a prototype of a class of stars known as *Cepheid variables* with differing average brightness and period. Many of them are in the range of parallax measurements so their distances are known and their average intrinsic brightness can be found from their average apparent brightness. In 1908 the Harvard astronomer Henrietta Leavitt (1868–1921; Figure 9.4(b)) found that the average intrinsic brightness and period were related; the relationship is shown in Figure 9.5 (Leavitt and Pickering, 1912).

From Figure 9.5 it will be seen that Cepheid variables can be very bright, up to about 30,000 times as bright as the Sun. They can therefore be seen at great distances, including in the outer regions of distant galaxies. Even if their apparent brightness is low, as long as the fluctuation in brightness can be seen the period can be determined, and hence the intrinsic average brightness. From the

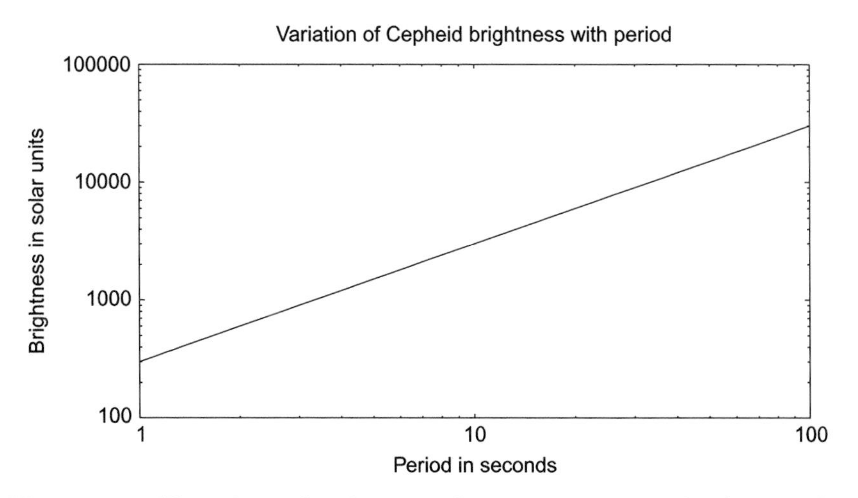

Figure 9.5. The relationship between the average intrinsic brightness of a Cepheid-variable star and its period.

apparent average brightness the distance can then be found. By the use of Cepheid-variables distances up to about 25 million pc can be determined. While this is a long way, the Universe stretches out much further.

The stated objective of this book is to show how we can determine the various properties of stars. These exist in galaxies beyond the range of Cepheid-variable distance determination and individual stars cannot be seen. Nevertheless, since these distant galaxies contain stars we will continue the narrative of how the distances of objects can be determined out to the furthest reaches of the Universe.

9.5. Distance Measurement Using Rotating Galaxies

To find distances beyond the range of Cepheid-variable determination we need an object of much greater brightness, the intrinsic brightness of which is known in some way. Spiral galaxies, similar to our galaxy, the Milky Way (Figure 3.2(a)), are extremely bright objects and the distances of many fall in the range of Cepheid-variable determination so, from their apparent brightness their intrinsic brightness can be found. The question is whether there is some property of a galaxy that relates to its intrinsic brightness, determinable from those galaxies within the Cepheid-variable range, which could be observed beyond the Cepheid-variable range.

From observations of stars within the Milky Way it can be deduced that our galaxy is spinning with a period of about 200 million years. Despite this being an extremely slow rotation, because the galaxy is so large, about 40,000 pc in average diameter, stars on opposite sides of the galaxy have relative speeds of order 1,000 km s^{-1}. If someone who was stationary with respect to the centre of the galaxy observed it edge on they would see light from one side red-shifted due to the Doppler effect as the stars were moving away and blue-shifted on the other side because stars were approaching. The amount of red or blue-shift would increase with distance from the centre. This situation is shown schematically in Figure 9.6.

Any spectral line in the light from the galaxy would be spread out due to the range of wavelength shifts from various parts of the

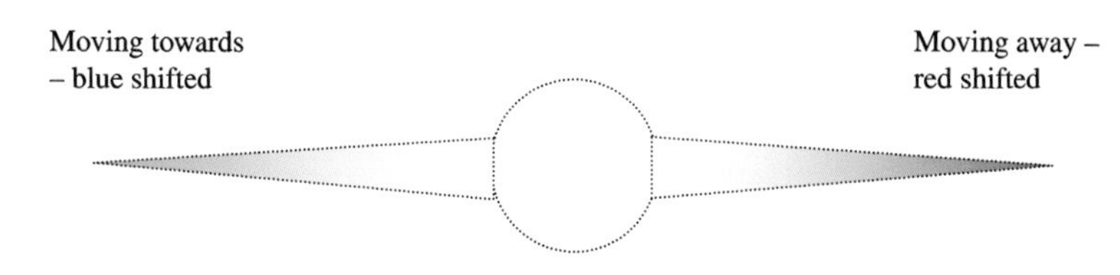

Figure 9.6. The spectral line shifts from a rotating spiral galaxy seen edge-on.

galaxy. The width of the spread would be independent of any overall motion of the galaxy relative to the observer. From observations of those spiral galaxies within the Cepheid-variable range it has been found that there is a relationship between the spread of spectral lines and the intrinsic brightness of the galaxy. The link involves the mass of the galaxy. The greater the mass the faster it spins and the greater is the number of stars and hence brightness. The connection between the spread of spectral lines and intrinsic brightness is well established and is known as the *Tully–Fisher relationship* (Tully and Fisher, 1977). By measuring the spread of the spectral lines of a distant spiral galaxy its intrinsic brightness can be estimated. From its apparent brightness its distance can then be estimated. In this way, distances can be measured out to about 600 million light years; beyond which distance the galaxy is too faint to give a good estimate of the spectral-line widths.

9.6. Distance Estimation from Type 1a Supernovae

We have already seen that stars, somewhat more massive than the Sun, will end their lives as energy-generating stars by exploding as supernovae. When they do this they become very bright objects, with energy output up to ten billion times that of the Sun and they are visible throughout the Universe. Although very bright they are of no use for distance estimation since their brightness varies from one supernova to another, depending on the mass of the exploding star. There are other kinds of supernovae that occur and the different types can be recognized by the characteristics of their spectra. One of these, a Type 1a supernova, has the characteristic that they all have the same peak brightness, which makes them useful tools for distance estimation.

The scenario that produces a Type 1a supernova is when there is a binary system in which one of the stars is a red giant and the other is a white dwarf. This would originally have been a binary system containing two main-sequence stars, both of which would eventually end up as white dwarfs but which evolved at different rates to give the red giant-white dwarf combination. Red giants shed material from time-to-time and this can be captured by the white dwarf. Thus the white dwarf steadily gains mass and when it reaches a critical mass known as the *Chandrasekhar limit* (Chandrasekhar, 1931), about 1.44 solar masses, it becomes unstable and explodes to give a Type 1a supernova. All Type 1a supernovae have the same characteristics because they come about in exactly the same way — they all have to reach the same critical mass.

Type 1a supernovae can be seen in very distant galaxies and because some occur in galaxies for which Cepheid-variable distances have been determined their maximum intrinsic brightness is known. Using Type 1a supernovae, distances of galaxies out to a distance of a billion parsecs can be determined; even so, we have still not yet reached the limits of the observable Universe.

Figure 9.7 shows the ranges of the various techniques that have been described. Outside the parallax region the boundaries move forward by finding ever brighter sources, the brightness of which are known because they also occur in a closer-in region. To reach the

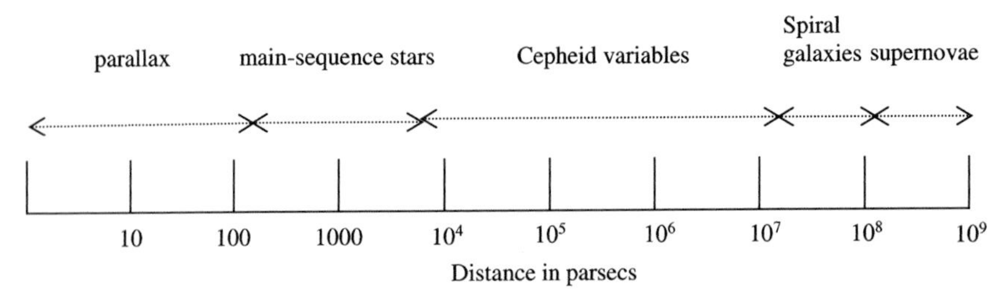

Figure 9.7. Techniques for measuring distances. Methods can be used below the limits shown, but not above.

edge of the observable Universe some new source, even brighter than a Type 1a supernova would be required.

Problems 9

9.1 **(a)** A main-sequence star of known absolute magnitude 5.7 is observed with an apparent magnitude of 19.1. What is its distance?

(b) A star is at a distance of 153 pc, as determined by parallax measurements. If its apparent magnitude is 6.4 then what is its absolute magnitude?

9.2 **(a)** A Cepheid variable is observed with a period of 3.0 s, indicating an absolute magnitude of –2.03. Its apparent magnitude is 20.4. What is its distance?

(b) At what distance would its apparent magnitude be 0?

9.3 The peak absolute magnitude of a Type 1a supernova is -19.3. One is observed in a distant galaxy with an apparent magnitude of 11.3. How far away is the galaxy?

Part 5

The General Properties of Stars

Chapter 10

Determining the Radii of Stars

10.1. The Radii of Main-Sequence Stars

All main-sequence stars are too far away to be seen as other than point sources of light. The way in which the radii of main-sequence stars, or indeed any other type of star, can be found is contained in equation (4.1), which we repeat here.

$$L = 4\pi\sigma R^2 T^4. \tag{10.1}$$

If the temperature and luminosity of a star are known then the radius can be determined.

The radius of a star that is determined is that of its photosphere, the source of the light that it emits. However, the substance of a star extends well beyond that limit, something that can be detected for the Sun when there is a solar eclipse (Figure 10.1). The photosphere is completely blocked out by the Moon and this enables the glowing atmosphere, known as the *solar corona* to be seen. The corona has a temperature of several million degrees, the reason for which has not been established. Despite this high temperature the coronal material is so diffuse that the light coming from it is swamped by that from the photosphere and, normally, it cannot be seen.

What is indicated by the theoretical curves shown in Figure 4.2 is that all main-sequence stars of a given mass end up at the same point on the main-sequence line in the H–R diagram, i.e. with the same temperature and luminosity and hence the same radius. If we

Figure 10.1. The solar corona seen during a total eclipse of the Sun.

take a star of one solar mass, then its luminosity is 3.828×10^{26} W and temperature 5,778 K. From (10.1)

$$R = \left(\frac{L}{4\pi\sigma}\right)^{1/2} \frac{1}{T^2} = \left(\frac{3.828 \times 10^{26}}{4\pi \times 5.67 \times 10^{-8}}\right)^{1/2} \frac{1}{5,778^2}$$
$$= 6.943 \times 10^8\,\text{m} = 694,300\,\text{km},$$

which is slightly different from the normally accepted value 695,700 km. This difference of values is of the same order as the variation to be expected within one stellar class. Not all G2V stars are *precisely* the same. Class G2V contains a small range of masses and measurements of temperature and luminosity are also prone to small errors. Another factor is that stars do change their properties slightly during their main-sequence lifetime.

The Table 10.1 gives mass, radius, luminosity and temperature for a selection of stellar classes. Mass, radius and luminosity are given in solar units and the small number of significant figures reflects the limited precision with which the parameters are known.

Table 10.1. Properties of some main-sequence stars.

Stellar class	Mass $M/M_\odot$	Radius $R/R_\odot$	Luminosity $L/L_\odot$	Temperature K
O5	60	12	1.4×10^6	53,000
B0	18	7.4	20,000	30,000
A0	3.2	3.5	80	10,800
F0	1.7	1.3	6	7,249
G0	1.1	1.05	1.26	5,920
K0	0.78	0.85	0.4	5,240
M0	0.47	0.63	0.063	3,920

10.2. The Radii of Giant Stars

Main-sequence stars are completely characterized by their spectral class, which determines their mass, radius, temperature and luminosity. They are more-or-less fixed in their characteristics for a long period of time in a small region of the H–R diagram on or near the main-sequence line. By contrast, a giant star can be anywhere in the region of point C in Figure 4.4 and the radius of a giant star cannot be determined from its spectral class.

There are some giant stars that are close enough for a parallax distance to be estimated and large enough for an image to be produced and from this information estimates can be made of their radii. One such star is Betelgeuse, a disk image of which was produced as early as 1920. An image produced in 2017 by ALMA (Atacama Large Millimetre/submillimetre Array) is shown in Figure 10.2. ALMA is a combination of 66 radio telescopes situated in the Atacama Desert in Chile. They are distributed over a wide area, many kilometres in extent, and working together they give a resolution equivalent to that of a single radio telescope several kilometres in diameter.

The distance of Betelgeuse is about 220 pc, with a possible error of 20%, and its observed angular diameter is between $0.042''$ and $0.056''$, depending on the observing wavelength. The estimated radius is of order $900R_\odot$ with a possible error up to 20%.

Figure 10.2. An image of Betelgeuse produced by ALMA (NASA).

Table 10.2. Characteristics of some giant stars.

Star name	**Spectral class**	**Distance (pc)**	**Radius ($R_\odot$)**
Aldebaran (α Tauri)	K5III	20.0	44.2
Arcturus (α Bootis)	K0III	11.3	25.4
Gacrux (γ Crusis)	M3.5III	27.2	84
Canopus (α Connae)	A9II	95	71

For any red giant for which the distance in parsecs, d, has been found the steps for finding the radius, R, from the apparent brightness, b and spectral class, giving temperature, T, are as follows:

From equation (9.5) $B = b\dfrac{d^2}{100}$.

From equation (9.4) $L = 400\pi B$.

From equation (10.1) $R = \left(\dfrac{L}{4\pi\sigma}\right)^{1/2} \dfrac{1}{T^2}$.

Table 10.2 gives the spectral class, distance and radii of some giant stars. Canopus is a *bright giant*, with mass about eight solar masses and a temperature of 6,998 K.

10.3. The Radii of White Dwarfs

All the considerations that were mentioned for the observational determination of the radii of giant stars can be applied to white dwarfs. If their distance, brightness and temperature have been determined then their radii can be found.

It turns out that from a theoretical point of view the radius of a white dwarf depends only on its mass. This will now be shown using some ideas from quantum mechanics but assuming that the speeds of all the particles involved are small compared with the speed of light, so that classical formulae for energy and momentum can be used rather than relativistic ones. The analysis for relativistic conditions is given later in the chapter.

10.3.1. *The Nature of White Dwarf Material*

An atom of normal matter consists of a tiny nucleus, containing positively-charged protons and neutral neutrons, with negatively-charged electrons occupying a comparatively large region, with radius of order 8×10^{-11} m for less massive atoms. The nucleons — protons and neutrons — have approximately the same mass, 1.67×10^{-27} kg, and the electron mass, 9.1×10^{-31} kg, is about 1/1840 of that of a nucleon. The most common ways that atoms bond together, to form either molecules or framework structures, is either by sharing some of their outer electrons or by one atom donating one or more of its electrons to a neighbouring atom. The mean densities of materials are of the same order as, but less than, the mean densities of the atoms forming them. For example, an atom with ten nucleons with a radius of 8×10^{-10} m would have mean density

$$\rho = \frac{10 \times 1.67 \times 10^{-27}}{\frac{4}{3}\pi\,(8 \times 10^{-11})^3} = 7.8 \times 10^3\,\text{kg m}^{-3},$$

a few times the density of water. However, the material of a white dwarf has ten million times the density of water, so it is clear that its material is not in atomic form. It is in a degenerate state, described in Appendix E, that we may think of as an intimate mixture of nucleons and electrons, not interacting and tightly packed together with no particular structure.

10.3.2. *Fermions*

Protons, neutrons and electrons are all *fermions* and a property of fermions is that they have *half-integral spin*, a description of an intrinsic angular momentum associated with them. The 'half' comes about because when quantum mechanics is applied to the electronic structure of atoms it turns out that all electrons must have angular momentum associated with their state that is a whole number of the units $\hbar = h/(2\pi)$, where h is Planck's constant (see equation (6.1)). However, the angular momentum associated with the spins of fermions is $^1/_2\hbar$ — hence 'half-integral spin'. Associated with their spins, fermions act like tiny magnets.

10.4. Basis of a Theoretical Approach to White Dwarf Structure

The purpose of this analysis is to find the relationship between the mass of a white dwarf and its radius. To do this we will find the total energy of the white dwarf — degeneracy kinetic energy plus gravitational potential energy — and then find the radius for which it is a minimum. This approach is based on the development of quantum mechanics in the 1920s, mainly by the German theoretical physicist, Werner Heisenberg (1901–1976; Figure 10.3(a)) and the Austrian, Erwin Schrödinger (1887–1961: Figure 10.3(b)).

10.4.1. *Degeneracy Kinetic Energy*

The degeneracy kinetic energy is given by equation (E.7), which is for particles of mass m Since electrons are much less massive than nucleons it is clear that only electrons contribute substantially to the degeneracy kinetic energy. A white dwarf consists primarily of carbon for which the number of electrons is one half of the number of nucleons, so we can take

$$N = \frac{M}{2m_N}, \tag{10.2}$$

(a) (b)

Figure 10.3. (a) Werner Heisenberg (German Federal Archives), and (b) Erwin Schrödinger.

where N is the number of electrons in the white dwarf, M its mass, r its radius and m_N is the average nucleon mass. In these terms the kinetic energy, derived from equation (E.6), is

$$E_K = \frac{3}{10}\left(\frac{9}{32}\right)^{2/3} \frac{\hbar^2}{\pi^{4/3}} \frac{1}{r^2 m}\left(\frac{M}{2m_N}\right)^{5/3}. \tag{10.3}$$

10.5. An Approximate Treatment of White Dwarf Structure

The total energy, E_T, of a white dwarf consists of two parts, the degenerate kinetic energy and the gravitational potential energy. Hence from (10.3) and (2.4), since a white dwarf will have close-to-uniform density,

$$E_T = \frac{3}{10}\left(\frac{9}{32}\right)^{2/3} \frac{\hbar^2}{\pi^{4/3}} \frac{1}{r^2 m}\left(\frac{M}{2m_N}\right)^{5/3} - \frac{3GM^2}{5r}. \tag{10.4}$$

The white dwarf will be stable if the radius is such that the total energy is a minimum, which will be true if $dE_T/dr = 0$.

Applying this condition gives

$$r = \left(\frac{9}{32}\right)^{\frac{2}{3}} \frac{\hbar^2}{Gm\pi^{\frac{4}{3}}} \left(\frac{1}{2m_N}\right)^{5/3} \frac{1}{M^{1/3}}. \tag{10.5}$$

Substituting for the various physical constants we find

$$r = 9.05 \times 10^{16}/M^{1/3},$$

where r is in metres if M is in kilograms. Thus the radius of a white dwarf with the mass of the Sun is 7.18×10^6 m or about 10% more than the radius of the Earth. It will be seen that the relationship gives the interesting result that the radius *decreases* with increasing mass of the white dwarf. It also suggests that there can be white dwarfs of any mass with any radii. This is not so, but to see why we have to introduce relativistic conditions that reveal that there is some limiting mass for a white dwarf at which its radius becomes vanishingly small.

10.6. A Relativistic Treatment

For very high densities the uncertainty in position of the electrons becomes very small. Consequently, from a consideration of the Heisenberg uncertainty principle, their momenta and kinetic energies become very large and we must resort to relativistic mechanics. Now, kinetic energy expressed as $p^2/2m$ in classical mechanics must be replaced by the corresponding relativistic expression $(p^2c^2 + m^2c^4)^{1/2} - mc^2$ so that the expression for the kinetic energy of a star of mass M and radius r, corresponding to equation (E.5), is

$${}_ME_K(r) = \frac{32\pi^2 r^3}{3h^3} \int_0^P \left\{ \left(p^2c^2 + m^2c^4\right)^{1/2} - mc^2 \right\} p^2 dp, \tag{10.6}$$

with P corresponding to $P_{\max}$ in (E.3). By a change of variable to $q = p/mc$ this is transformed to

$${}_ME_K(r) = Ar^3 \int_0^B \left\{ \left(1 + q^2\right)^{1/2} - 1 \right\} q^2 dq, \tag{10.7}$$

where $A = \frac{32\pi^2 m^4 c^5}{3h^3}$ and $B = \frac{h}{mc}\left(\frac{9}{64\pi^2 m_N}\right)^{1/3} \frac{M^{1/3}}{r}$.

The total energy of the star is thus

$$E_T = {}_M E_K(r) - \frac{3GM^2}{5r}, \tag{10.8}$$

with the condition for stability again $dE_T/dr = 0$, corresponding to a minimum energy. If dE_T/dr is positive for all values of r then the star will shrink without limit, since this continuously reduces the energy. The simplest way to explore the conditions for stability, or its lack, is numerically. The finite-difference approximation

$$\frac{d\,({}_M E_K(r))}{dr} = \frac{{}_M E_K(r+\delta) - {}_M E_K(r-\delta)}{2\delta},$$

with integrals evaluated by numerical quadrature enables the derivative of the first term on the right-hand side of (10.8) to be easily found for any combination of M and r.

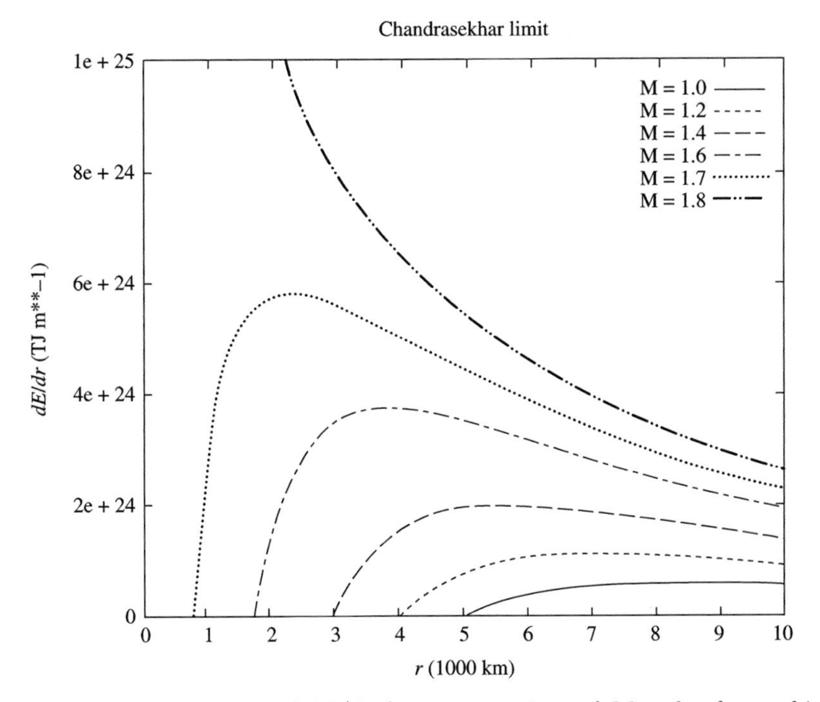

Figure 10.4. The values of dE/dr for various values of M and r for a white dwarf. Above about $M = 1.7\,M_\odot$ no radius gives stability. This corresponds to the Chandrasekhar limit.

Figure 10.4 shows the results of such a calculation. For a solar mass the radius for stability is about 5,000 km — less than the radius of the Earth and less than the value found using the classical expression for kinetic energy. For a mass of 1.7 $M_\odot$ the stability radius is about 800 km but for a mass of 1.8 $M_\odot$ it can be seen that dE_T/dr is always positive. The greatest mass at which no stable configuration can exist is the *Chandrasekhar limit.* Here we have found it to be somewhat over 1.7 $M_\odot$ but the value usually quoted is 1.44 $M_\odot$. In his original treatment Chandrasekhar considered factors that are not included here.

10.7. Neutron Stars and Black Holes

Within a collapsing star consisting of degenerate material of mass exceeding the Chandrasekhar limit, the pressure becomes so great that the electrons in the star combine with the protons in the ions to form neutrons so that the star becomes a *neutron star.* Neutrons are also spin-$^1/_2$ particles, fermions, so they are able to generate a *neutron degeneracy pressure.* Since they are much more massive than electrons then, according to (E.8), they must be much more highly compressed before they exert the pressure necessary to resist gravity and, consequently, neutron stars are even smaller than white dwarfs. Investigating the size of a neutron star can be done by the same sort of analysis as has been given above with minor modifications. The number of degenerate particles is now equal to the number of nucleons (not one-half as many as in the electron case) since all the particles are now neutrons. The other change is that the neutron mass must be used where previously the electron mass appeared. The derivative of (10.8) is evaluated to find the stability conditions but now with

$$A = \frac{32\pi^2 m_n^4 c^5}{3h^3} \quad \text{and} \quad B = \frac{h}{mc}\left(\frac{9}{32\pi^2 m_n}\right)^{1/3}\frac{M^{1/3}}{r},$$

where m_n is the mass if the neutron. Calculations for the neutron star are shown in Figure 10.5. A neutron star of mass 2 $M_\odot$ has a radius of about 9 km and this falls to less than 3 km for a mass of 6 $M_\odot$.

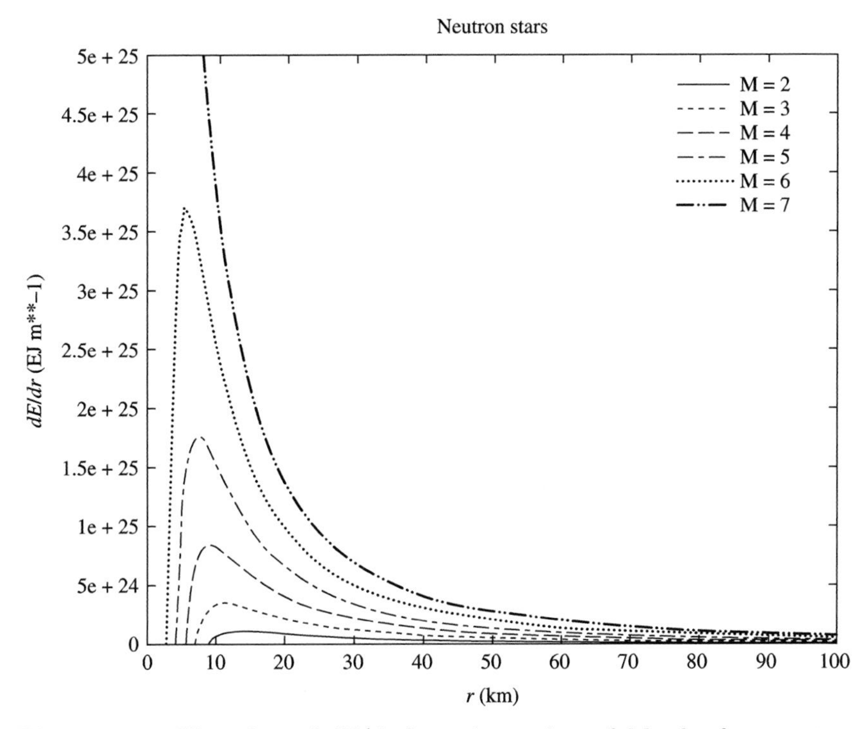

Figure 10.5. The values of dE/dr for various values of $M$and r for a neutron star. The limiting mass for a neutron star is indicated as between 6 and $7\,M_{\odot}$.

For a mass of $7\,M_{\odot}$ there is no stable configuration and the star will collapse without limit to a *black hole.* Most published estimates of the maximum mass of a neutron star are $3\,M_{\odot}$ or so but some are as high as $6\,M_{\odot}$, which is suggested by the present analysis.

Problems 10

10.1 A giant star, at a distance of 310 pc and with temperature 3,900 K, has a measured brightness of 9.30×10^{-10} W m^{-2}. Determine its radius.

10.2 Using the classical approximation find the radius of a white dwarf of mass $0.8\,M_{\odot}$. What is its mean density? Find the relationship between density and mass for a classical white dwarf.

10.3 A neutron star has the density of a neutron, which has a radius of 10^{-15} m. What is the spin period of a neutron star of mass $2\,M_{\odot}$ if the centripetal acceleration at the equator is 0.1 of the gravitational field at the surface?

Chapter 11

Determining the Masses of Stars

11.1. General Comments

It has been inferred, for example in Table 4.1, that for main-sequence stars the spectral class depends on the mass, so by observing the spectral class of any main-sequence star the mass can be estimated even if we know nothing more about it. However, what we are concerned with here is determining the masses of stars from observations, and these can be any kind of stars, main-sequence, dwarf or giant — and, as we shall see, even invisible stars in some favourable circumstances.

An astronomical body reveals its mass through its gravitational influence on other bodies so it is evident that the mass of an isolated body cannot be determined by observational measurements. Fortunately, most stars are members of binary systems; there are about equal numbers of binary systems and isolated field stars, so two-thirds of all stars are members of a binary system. It is by analysis of the motions of the two stars of a binary system that it is possible to estimate their masses.

11.2. Kepler's Laws

The outstanding observational astronomer of the 16th century was the Danish nobleman, Tycho Brahe (1546–1601; Figure 11.1). Before the invention of the telescope he used large-scale line-of-sight

Figure 11.1. Tycho Brahe.

instruments situated in his observatory, Uraniborg, situated on the island of Hven between Denmark and Sweden, to plot the motions of planets with much greater precision than had ever been done previously. However, his qualities as an astronomer outshone his qualities as a human being. He was querulous in the extreme and had part of his nose sliced off in a student duel; thereafter he wore a prosthetic nose made of gold and silver. He was granted an estate on Hven by Frederick II, the king of Denmark and Norway, but showed very little gratitude for the king's benevolence and also treated the tenants on the estate very badly. In 1597, Christian IV succeeded to the throne and shortly afterwards Brahe was banished from Hven. He was invited by Rudolph II, the king of Bohemia, to go to Prague as court astronomer and from 1600 to his death in 1601 he was assisted by a brilliant German mathematician and astronomer, Johannes Kepler (1571–1630; Figure 7.6).

A star plus a single planet would be like a stellar binary system but with the property that the ratio of the masses of the two bodies is large. With access to Tycho Brahe's extensive observations after his death, Kepler set about analyzing the motions of the planets. Eventually he arrived at three laws of planetary motion,

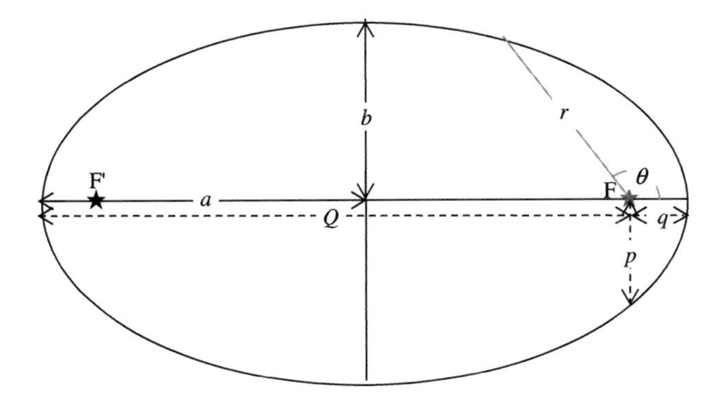

Figure 11.2. The geometry of an ellipse.

which were

(1) Planets have elliptical orbits with the Sun at one focus.
(2) The line connecting the planet to the Sun sweeps out equal areas in equal time.
(3) The square of the orbital period is proportional to the cube of the semi-major axis of the orbit.

The law of main interest in the present context is the first for, although binary systems do not normally have extreme ratios of stellar masses, the orbit of each star is an ellipse, the characteristics of ellipses and elliptical orbits are now described.

11.3. The Characteristics of an Ellipse

The general appearance of an ellipse is illustrated in Figure 11.2. The point that Kepler identified as the *focus* is the point F, a is the *semi-major axis* and b the *semi-minor axis* of the ellipse. These quantities are related through a quantity known as the *eccentricity*, e, of the ellipse by

$$b^2 = a^2(1 - e^2). \tag{11.1}$$

The eccentricity must satisfy $0 \leq e < 1$; when $e = 0$ then $a = b$ and the ellipse has become a circle. There is another focus, F′, which is not occupied. Both foci are at a distance ae from the centre of the ellipse.

For considering orbital motions, an ellipse is best described in polar coordinates (r, θ), as seen in the figure, by

$$r = \frac{a(1-e^2)}{1+e\cos\theta}. \tag{11.2}$$

The *semi-latus rectum* of the ellipse, p, is shown in Figure 11.2 and corresponds to the value of r when $\theta = \pm\pi/2$, i.e. $\cos\theta = 0$, Although it will not feature in our analysis of binary systems it is an important quantity because it defines the angular momentum contained in a system. The semi-latus rectum is given by

$$p = a(1-e^2). \tag{11.3}$$

The closest distance of the orbit to F is when $\theta = 0$ and this distance, q, is known as the *perihelion distance* or sometimes just *perihelion* when the body at F is the Sun, but if the body is a general star then the term *periastron* is used. Similarly Q, the furthest distance from F, corresponding to $\theta = \pi$, is the *aphelion distance* (or *aphelion*) when the Sun is at F or *apastron* if a general star is at F. From equation (11.2) with the appropriate values of $\cos\theta$

$$q = a(1-e), \tag{11.4a}$$

and

$$Q = a(1+e). \tag{11.4b}$$

11.4. The Centre of Mass and the Orbits of Binary Stars

Consider a system of N particles for which the ith particle has coordinates (x_i, y_i, z_i) and mass m_i. The centre of mass of the system is at (X, Y, Z) where

$$X = \frac{\sum_{i=1}^{N} m_i x_i}{\sum_{i=1}^{N} m_i}, \quad Y = \frac{\sum_{i=1}^{N} m_i y_i}{\sum_{i=1}^{N} m_i}, \quad Z = \frac{\sum_{i=1}^{N} m_i z_i}{\sum_{i=1}^{N} m_i}. \tag{11.5}$$

Such a system, placed in a uniform gravitational field, would be in balance about a fulcrum at position (X, Y, Z).

For a binary system there are just two bodies and the centre of mass is situated on the line joining them (Figure 11.3). If the stars are at S_1 and S_2 and have masses M_1 and M_2 respectively then, with the centre of mass at C,

$$\frac{S_1\mathrm{C}}{S_2\mathrm{C}} = \frac{M_2}{M_1}. \tag{11.6}$$

From (11.6) it is clear that the centre of mass is closer to the more massive of the two stars.

Binary systems vary greatly in the separation of the two stars. If the stars are close then there are mutual tidal effects that tend to circularize the orbits. However, for large separations the orbits will usually be general ellipses. In Figure 11.4 the orbits of the two stars of a binary system are shown with corresponding positions at three times. The centre of mass is also a focus of each of the orbits, corresponding to the point F in Figure 11.2.

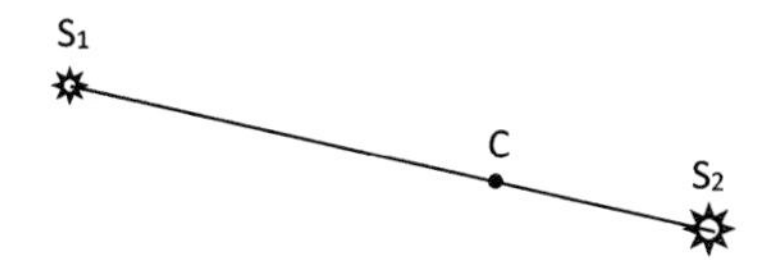

Figure 11.3. Two stars and their centre of mass.

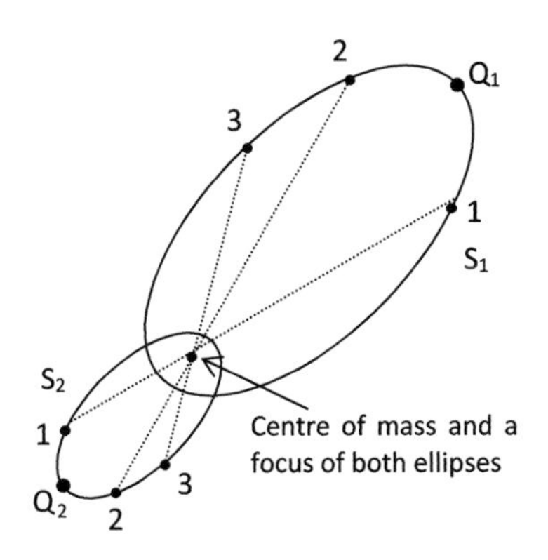

Figure 11.4. The elliptical orbits of the two stars with corresponding positions.

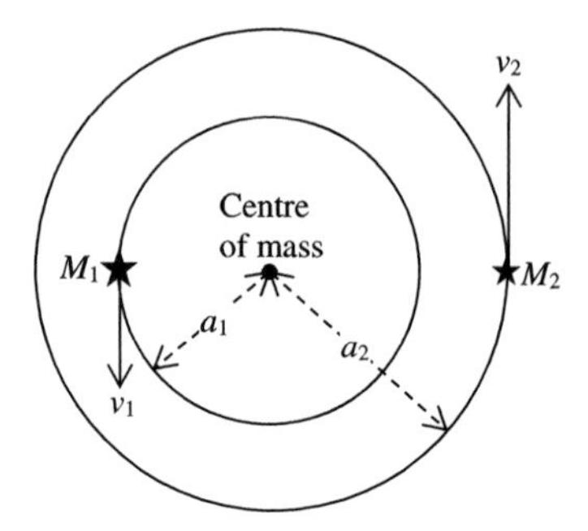

Figure 11.5. The circular orbits of two stars around the centre of mass.

11.5. The Mathematics of Binary Star Orbits

To keep things simple, here we will just show how to find the masses of the stars of a binary system for circular orbits and then quote the corresponding results for elliptical orbits. Figure 11.5 shows the circular orbits of the two stars of masses M_1 and M_2 in orbits of radius a_1 and a_2 at speeds v_1 and v_2.

The period, P, equals the circumference of an orbit divided by the speed, so that

$$P = \frac{2\pi a_1}{v_1} = \frac{2\pi a_2}{v_2}. \tag{11.7}$$

The centrally directed force on each star must equal the centrifugal force, giving

$$\frac{GM_1M_2}{a^2} = \frac{M_1v_1^2}{a_1}, \tag{11.8}$$

where $a = a_1 + a_2$. Eliminating v_1 from (11.7) and (11.8)

$$P^2 = \frac{4\pi^2 a_1 a^2}{GM_2}. \tag{11.9}$$

From (11.6), given that the overall size of each orbit is proportional to its radius,

$$\frac{a_1}{M_2} = \frac{a_2}{M_1} = \frac{a_1 + a_2}{M_2 + M_1} = \frac{a}{M}, \tag{11.10}$$

where $M = M_1 + M_2$. Substituting in (11.9)

$$P^2 = \frac{4\pi^2 a^3}{GM}. \qquad (11.11)$$

This equation is an expression of Kepler's third law in the context of binary systems, that the square of the period is proportional to the cube of the semi-major axis. Equation (11.11) is also true for elliptical orbits where $a = a_1 + a_2$ and a_1 and a_2 are the semi-major axes of the two orbits.

11.6. Determining the Masses of Stars in Binary Systems

Binary systems can be divided into four main types, which are:

Wide or *visual binaries*, where the stars are far enough apart for them to be seen individually and for their projected paths on the night sky to be determined.

Close or *spectroscopic binaries*, where even with a telescope the individual stars cannot be resolved but which can be recognized as binary systems by looking at their combined spectra.

Eclipsing binary, where the line of sight is in the plane of the stellar orbits and each star periodically passes in front of its companion.

Astrometric binaries, where a single star is seen orbiting about a point in space but without the companion star being visible.

We now consider what information about stellar masses can be determined for each type of system.

11.6.1. *Wide Binaries*

We shall first assume that the line of sight is perpendicular to the plane of the orbits so that the precise shapes of the orbits can be determined. The second assumption is that the distance, D, of the system from the Earth is known. It could be known through any of the mechanisms described in Chapters 7, 8 and 9 but most easily if one of the components is a main-sequence star, the apparent brightness of which can be measured.

The first measurements required are the maximum angular separations, α_1 and α_2, of the positions for each of the two stars. From this

$$a = a_1 + a_2 = \alpha_1 D/2 + \alpha_2 D/2, \tag{11.12}$$

since the maximum separations, from periastron to apastron, are $2a_1$ and $2a_2$.

The period can be measured quite straightforwardly and then from equation (11.11)

$$M = \frac{4\pi^2 a^3}{GP^2}. \tag{11.13}$$

To find the individual masses we use (11.10) so that

$$M_1 = \frac{a_2}{a} M \text{ and } M_2 = \frac{a_1}{a} M. \tag{11.14}$$

Now it is extremely unlikely that the line of sight will be perpendicular to the plane of the orbits. If the line of sight makes an angle i with the semi-major axes — the line Q_1Q_2 in Figure 11.4 — then a_1 and a_2, and hence a, will appear foreshortened by a factor $\sin(i)$. Then the deduced value of the sum of the masses found from (11.13) will be

$$M' = M \sin^3 i, \tag{11.15}$$

and the values of M_1 and M_2 will be underestimated by a similar factor. Since i is unknown we can only find lower-bound estimates for the masses of the individual stars.

11.6.2. *Spectroscopic and Eclipsing Binaries*

Here we will initially make two assumptions. The first is that the line of sight is contained in the plane of the stellar orbits and the second is that the orbits are circular. In fact, as previously indicated, the second assumption is very likely to be true, or nearly so, for a spectroscopic binary since the tidal effects between very close stars in a binary system tend to round-off their orbits.

Although the motions of the individual stars cannot be seen the effects of their motions can be detected through Doppler shifts of spectral lines. This is illustrated in Figure 11.6. The two stars may be of different spectral classes but they will have some spectral lines in common. The figure shows how a line in the green part of the spectrum of each of the stars is Doppler shifted as the stars move round in their orbits. When their motions are transverse with respect to the line of sight the spectral lines are not shifted. When a star moves towards the observer the line is blue-shifted and when away from the observer the line is red-shifted. The maximum difference of the wavelength shift in one orbit for an orbital speed v is

$$d\lambda_{diff} = \lambda \frac{2v}{c} \text{ or } v = c\frac{d\lambda_{diff}}{2\lambda}, \tag{11.16}$$

where c is the speed of light.

If the whole binary system is moving away from or towards the observer then the wavelength with the stars moving transversely will be very slightly shifted from the laboratory value but it will make an infinitesimal change to an observed $d\lambda_{diff}$ and the estimated value of v.

The observed quantities are the period P, and two speeds, v_1 and v_2 obtained using equation (11.16) from the corresponding values of λ_{max}. The steps in finding the masses of the stars are as follows:

(i) From (11.7) $a_1 = \frac{Pv_1}{2\pi}$ and $a_2 = \frac{Pv_2}{2\pi}$. This gives $a = a_1 + a_2$,
(ii) The sum of the stellar masses is found from (11.13),
(iii) The orbital speeds and stellar masses are related by $M_1v_1 = M_2v_2$. The simplest way to see this relationship is to consider the binary system as isolated so its linear momentum is unchanged. Since the velocities of the stars are in opposite directions then $M_1\mathbf{v}_1 + M_2\mathbf{v}_2 = \mathbf{0}$, where $\mathbf{0}$ is a *null vector* representing no change in the vector quantity, momentum,
(iv) From $M = M_1 + M_2$ and $M_1v_1 = M_2v_2$ it follows that

$$M_1 = M\frac{v_2}{v_1 + v_2} \text{ and } M_2 = M\frac{v_1}{v_1 + v_2}. \tag{11.17}$$

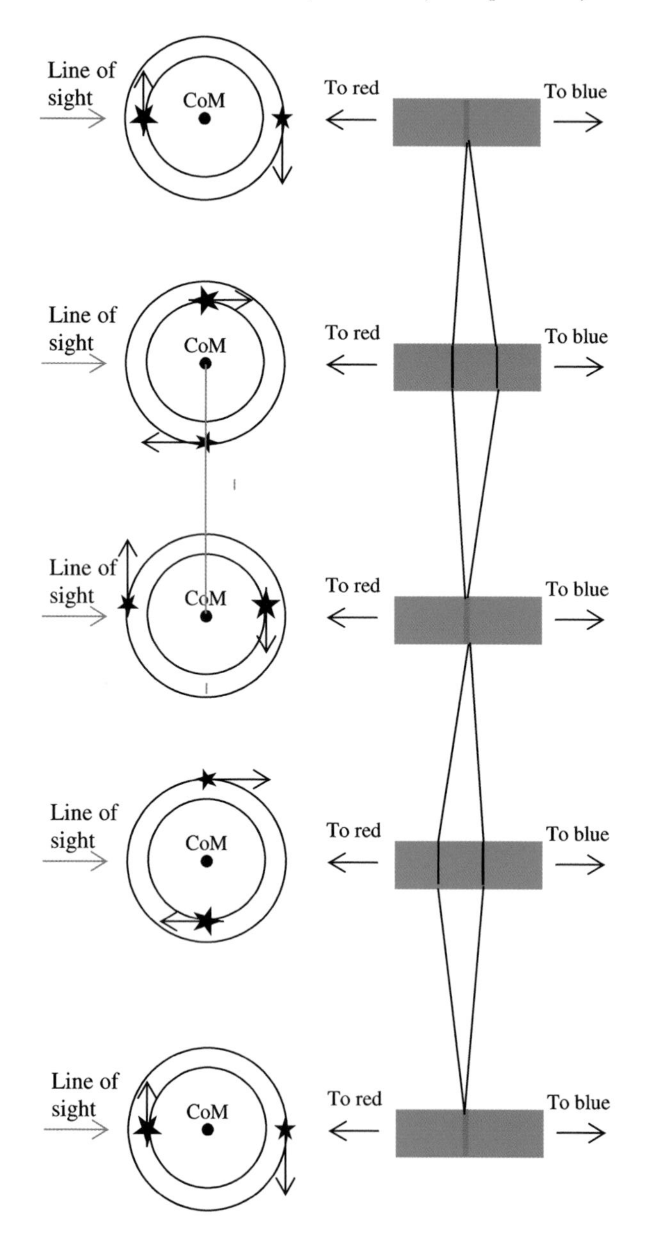

Figure 11.6. The shift of a spectral line due to stellar orbital motions.

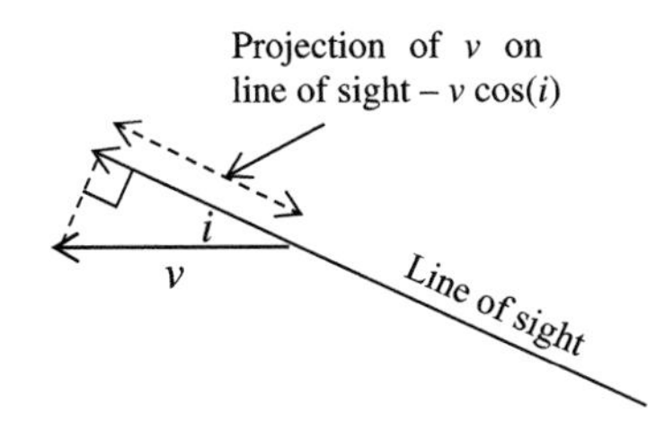

Figure 11.7. The projection of v on the line of sight.

Again there is the problem that the line of sight will not be in the orbital plane — if it were then the system would be an eclipsing binary, which can be recognized. If the line of sight makes an angle i with the orbital plane then the radial speed appears to be reduced by a factor $\cos(i)$ (see Figure 11.7). Hence the deduced values of a_1 and a_2, and also of a, as found from (11.7), will be too small by a factor $\cos(i)$ and M, found from (11.3), will be too small by a factor $\cos^3 i$. Finally the individual masses, as found from (11.17) will also be too small by the same factor.

An eclipsing binary is a special case of a spectroscopic binary where the angle $i = 0$ so there is no error in the mass estimates due to the inclination of the line of sight. Although the stars cannot be resolved, in addition to the variations in the wavelengths of spectral lines there are also variations in brightness that reveal the nature of what is being observed. The first variable star observed by John Goodricke in 1782 was Algol (β-Persei) and he explained the variation in its brightness as the consequence of it being an eclipsing binary star. The pair consisted of one very bright star with the other slightly smaller but much dimmer. The sequence of brightness is shown and explained in Figure 11.8. In position A the observer is seeing the light from both stars, giving maximum brightness. In position B the dimmer star is blocking out most of the light from the bright star and there is a large fall in brightness. In position C the dimmer star has moved to the other side of the bright one so both stars are delivering light to the observer and again there is maximum brightness. Finally, at position D the bright star is blocking out the light from the dimmer one so there is a small diminution in brightness.

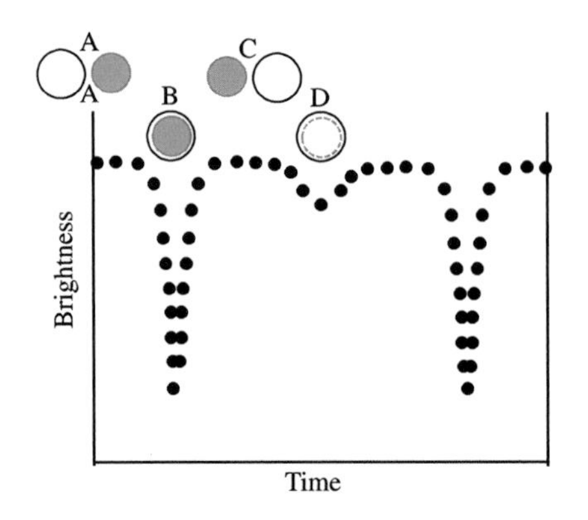

Figure 11.8. The light curve for an Algol-type eclipsing binary system.

Although individual radii of stars cannot be found from observations of eclipsing binaries it is possible to determine the ratio of the radii — as long as the eclipse is total so that when eclipsed the boundaries of the bodies do not overlap. If the brighter body has brightness b_1 and radius R_1 and the corresponding quantities for the dimmer body are b_2 and R_2 then the brightness seen at positions A, B and D are

$$b_{\rm A} = b_1 + b_2, \tag{11.18a}$$

$$b_{\rm B} = \left(1 - \frac{R_2^2}{R_1^2}\right) b_1 + b_2, \tag{11.18b}$$

$$b_{\rm C} = b_1. \tag{11.18c}$$

From (11.18a) and (11.18c) both b_1 and b_2 can be found and then from (11.18b) the ratio R_2/R_1. If one of the stars is main-sequence so that its radius is known from its observed spectral class then an estimate can be found for the radius of the other star.

11.6.3. *Astrometric Binary System*

In an astrometric binary system, where the orbit of just one component is seen, it is not, in general, possible to determine the mass of either component, However, if the distance of the binary is known we can estimate a_1, the semi-major axis of its orbit, and if the visible star is on the main sequence then, from its spectral class, we can estimate its mass, M_1. In addition, from observations of the orbit we can find the period, P. Given M_1, a_1 and P we can obtain information about the unseen star.

From (11.11)

$$M = \frac{4\pi^2 a^3}{GP^2}, \quad \text{or} \tag{11.19}$$

$$M_2 = \frac{4\pi^2 a^3}{GP^2} - M_1. \tag{11.20}$$

From equation (11.6)

$$M_2 = \frac{M_1 a_1}{a_2} = \frac{M_1 a_1}{a - a_1}. \tag{11.21}$$

Substituting for M_2 from equation (11.21) into equation (11.20) and rearranging gives

$$\frac{4\pi^2}{GP^2} a^3 - \frac{4\pi^2 a_1}{GP^2} a^2 - M_1 = 0. \tag{11.22}$$

Equation (11.22) is a cubic equation for determining a that, in general, has three roots. All three may be real or one may be real and the other two imaginary. There are standard methods of solving cubic equations but we will illustrate one particular solution using a graphic approach. Once a value of a has been determined then from a_1 we will know a_2 and hence, from (11.6), M_2.

The example taken is $M_1 = M_\odot = 1.989 \times 10^{30}\,\text{kg}$, $a_1 = 20\,\text{au} = 2.992 \times 10^{12}\,\text{m}$ and $P = 100\,\text{yr} = 3.156 \times 10^9\,\text{s}$. With these values equation (11.22) becomes

$$f(a) = 5.942 \times 10^{-8} a^3 - 1.778 \times 10^5\, a^2 - 1.989 \times 10^{30} = 0. \tag{11.23}$$

A plot of $f(a)$ as a function of a is shown in Figure 11.9. Although a must be positive the plot is from sufficiently large negative a to large positive a to be sure that the form of the curve, i.e. going towards $+\infty$ for large positive a and towards $-\infty$ for large negative

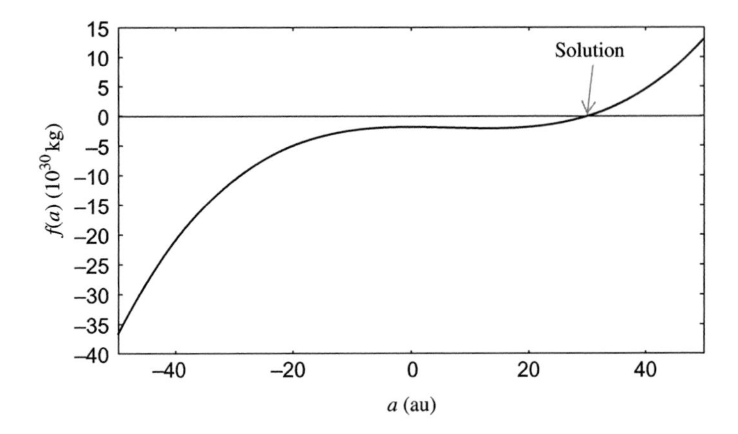

Figure 11.9. The solution of equation (11.23).

a, indicates that the region for either one or three real solutions has been straddled.

The only one real solution; by examining the numerical output of the computer program giving $f(a)$ this is found to be $a = 30.64$ au thus giving $a_2 = 30.64 - 20 = 10.64\,\text{au}$. Now from (11.6)

$$M_2 = \frac{a_1}{a_2} M_1 = \frac{20}{10.64} M_\odot = 1.88\, M_\odot.$$

As a check we look at the two sides of equation (11.19). The left-hand side is $M = M_1 + M_2 = 2.88\, M_\odot = 5.72 \times 10^{30}\,\text{kg}$. The right-hand side is

$$\frac{4\pi a^3}{GP^2} = \frac{4\pi^2 \times (30.64 \times 1.496 \times 10^{11})^3}{6.674 \times 10^{-11} \times (3.156 \times 10^9)^2} = 5.72 \times 10^{30}\ \text{kg}.$$

For this, as for all the other methods of estimating masses except for using eclipsing binaries, there will be an obliquity factor that will affect the value of a_1 and hence the deduced value of M_2.

Problems 11

11.1 A visual binary system is known to be at a distance of 159 pc. The maximum angular separation of positions for the two stars are respectively $0.00432''$ and $0.00392''$. The period of the orbit is 1.092 years. What are the masses of the two stars?

11.2 An iron spectral line, of laboratory wavelength 400.1664 nm, has a maximum wavelength difference of 0.0488 nm for one component of a spectroscopic binary system and 0.0344 nm for the other component. The period of the orbit is 200 days. What are the masses of the two stars? (Assume line of sight is in the plane of circular orbits).

11.3 One member of an eclipsing binary system is a main-sequence star of known radius 1.15 $R_{\boldsymbol{\phi}}$. The other star is of larger radius. When the stars are separated the brightness is taken as 1.0 in arbitrary units. When the main-sequence star is totally eclipsed by the larger star the brightness is 0.933 units and when the smaller star transits the disk of the larger one the brightness is 0.945. Estimate the radius of the large star in solar units.

11.4 Find the mass of the invisible star of an astrometric binary system if the visible star has mass 1.15 $M_{\odot}$, in an orbit of radius 20 au and period 70 years. The solution section provides a program, written in FORTRAN to perform this calculation.

Chapter 12

Other Stars and Star-like Objects

There are numbers of celestial objects that can be observed, or otherwise detected, that are either stars in unusual states or that resemble stars although they are not stars. Three such types of object are briefly described.

12.1. Pulsars

In 1967 Jocelyn Bell Burnell (b.1943, Figure 12.1(b)), a research student of the Cambridge astronomer Antony Hewish (b. 1924, Figure 12.1(a), Nobel Prize in Physics, 1974). looking at the output of a radio telescope they had constructed, noticed what she called a 'piece of scuff', which turned out to be pulsed radio signals at short and very regular intervals of time (Hewish *et al.*, 1968). The suspicion that they might be man-made and of terrestrial origin was ruled out when the source was found to keep sidereal time. The idea was floated that the regularity might indicate that they originated from some extra-terrestrial civilization, an idea picked up by the press, and for a time the signals were referred to as LGM — little green men. When a second source of the same kind was discovered it was clear that this was a purely astronomical phenomenon and the sources were then referred to as *pulsars*.

The source of a pulsar is a neutron star that is spinning very quickly and emitting a beam of radio-frequency electromagnetic radiation along the magnetic axis of a strong dipole-like magnetic

Figure 12.1. The pulsar people (a) Anthony Hewish (b) Joclyn Bell-Burnell.

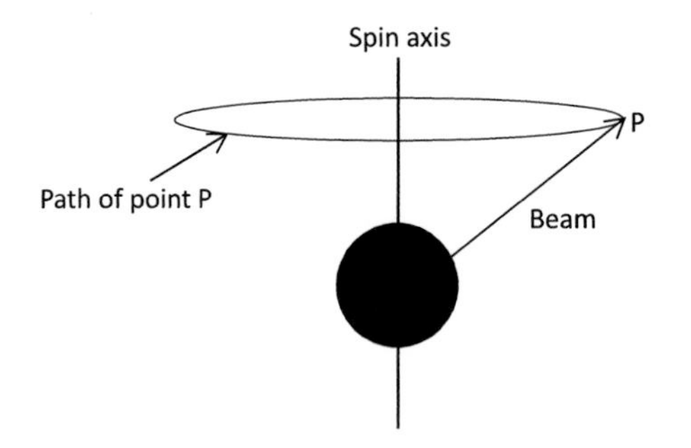

Figure 12.2. The relationship of the spin axis to direction of beam for a pulsar.

field. Since the spin axis does not correspond to the direction of the beam, the rotating beam traces a conical surface (Figure 12.2) and if it sweeps over the Earth at one point of each revolution then a stream of pulses will be detected. For different pulsars they can be at any interval in the range from seconds to milliseconds.

The Crab Nebula (Figure 12.3) represents the outflow from a supernova that was recorded in China and the Middle East in 1064. In 1988 a pulsar, of period 33 ms was discovered at its centre, the

Figure 12.3. The Crab Nebula with the Crab Pulsar at its centre (NASA).

first time a pulsar, i.e. a neutron star, had been directly associated with a supernova.

12.2. Quasars

Early in the 20th century it was realised that faint fuzzy objects seen in telescopes were distant galaxies but some of these were emitting vast amounts of electromagnetic energy covering the complete range from gamma rays to radio waves. In most cases no sources for this energy could be detected although very faint star-like objects were detected as likely sources in some galaxies. The observed spectral lines were initially unexplained until eventually they were interpreted as normal hydrogen lines with huge redshifts, up to 37%. If interpreted as Doppler shifts such redshifts would indicate large recession speeds, up to a considerable fraction of the speed of light. These sources became known as *quasars* — quasi-stellar objects — and their nature was a mystery for many years. The quasar 3C 273, seen in Figure 12.4, would have to have had a recession speed of 47,000 $\mathrm{km\,s^{-1}}$ to explain the observed redshifts.

Since quasars were associated with very distant galaxies then, to explain the brightness of the sources over the whole range of

Figure 12.4. The quasar 3C 273 (NASA/HST).

the electromagnetic spectrum, it was clear that their rate of energy output had to dwarf that of any conceivable star — in the case of 3C 273 about four trillion (4×10^{12}) times greater than that of the Sun.

Eventually in the 1960s and 1970s a plausible explanation for the properties of quasars was forthcoming. It is now known that at the centre of many galaxies, including our own, there exist massive black holes. Surrounding these black holes are accretion disks, the material of which is spiralling into the black holes in a process in which a high proportion of the absorbed mass is being converted into energy. Up to 1,000 solar masses per year can be absorbed by a black hole in this way. The redshift is not due to recession but due to the huge gravitational field within which the energy is being produced. A gravitational redshift is what is predicted by Einstein's General Theory of Relativity.

12.3. Wolf–Rayet Stars

These are massive stars at a late stage of their evolution when most of the hydrogen has been lost from their outer regions and where the main spectral signatures come from helium, carbon, nitrogen and oxygen. They tend to show strong broad emission lines from helium and predominant lines from one of carbon, nitrogen or oxygen. Depending on the dominant emission lines, other than those of helium, they are designated as WC, WN or WO stars, the terminal letter indicating the source element giving the dominant emission lines. They are also characterised by very strong stellar winds that are responsible for the loss of outer hydrogen.

Wolf–Rayet stars have very high temperatures — from about 30,000 to 200,000 K and corresponding high luminosities from thousands to over a million times that of the Sun.

Problem 12

12.1 The brightest quasar, 3C 273, has an apparent magnitude 12.8 and is at a distance estimated to be 7.49×10^8 pc.

(i) What is its absolute magnitude?

(ii) What is its luminosity in solar luminosity units? (The absolute magnitude of the Sun is 4.83)

Part 6

Exoplanets

Chapter 13

Planets About Other Stars

Now we are considering another interpretation of the title of this book, where 'about' has a meaning associated with location, i.e. 'around' or 'surrounding'. One of the great questions of science has been 'Are we alone?' or 'Is there other life in the Universe, be it intelligent or otherwise?' Effort is being expended to determine whether or not life exists, or has existed, in various locations within the Solar System, such as on Mars or in the ocean thought to exist below the surface of Saturn's satellite, Enceladus. Another line of approach has been to look for planets around other stars — *exoplanets* — and this approach has succeeded. For those interested in finding intelligent life the requirement is that of finding Earth-like exoplanets that has, or has had, liquid water, an environment within which evolutionary processes could produce complex life forms. If such exoplanets could be found then there would be the problem of communication — possibly insoluble.

However, whether or not exoplanets existed, whatever their form, was an interesting question in its own right and, as we shall discover, they certainly do.

13.1. Planets Around Pulsars

If we consider a star with a single planet then it is similar to a binary system but with the difference that the ratio of the mass of the star to that of the planet is very large. The bodies would each orbit around the centre of mass but the motion of the star would be on a much smaller scale than that of the planet, which would be invisible. Thus we have the situation of an astrometric binary system, but one in which the motion of the visible component is very restricted and hence difficult to measure precisely by Doppler-shift measurements.

In Section 12.1, the regularity of the pulses from pulsars was described. A very small and high-density object like a neutron star can rotate very rapidly, with a spin period down to milliseconds, without disruption. If it is detected as a pulsar then it acts like a very high-precision timekeeper.

In Figure 13.1, we show a pulsar moving around a circular orbit with an observer in the plane of the orbit, but at a considerable distance. If the pulsar-observer distance were constant then the observer would record the pulses at a constant frequency, say n. However, if the pulsar were moving relative to the observer then the frequency would change to

$$n' = \frac{1}{1 + \frac{v}{c}} n, \tag{13.1}$$

where c is the speed of light and v is the component of the relative velocity along the line of sight, with v positive for motion away from the observer. This is simply the Doppler-effect equation.

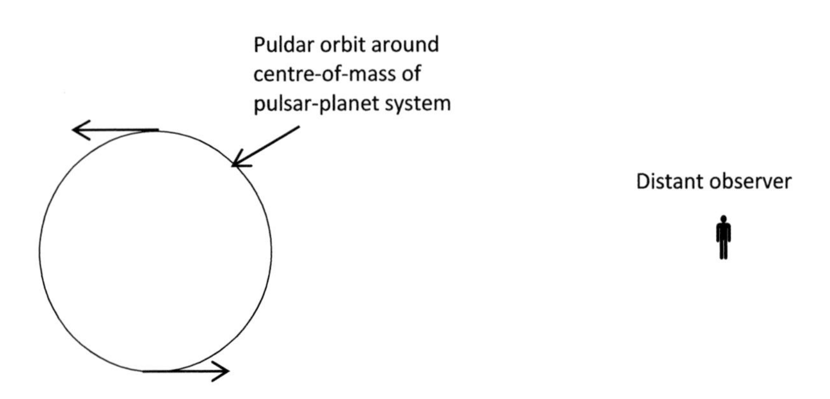

Figure 13.1. An orbiting pulsar and a distant observer.

The first confirmed exoplanets were discovered in 1992 orbiting the millisecond pulsar PSR 1257 + 12 (Wolszczan and Frail, 1992). A single planet would have given a periodic variation of the frequency of the pulses but there were two planets initially discovered, of masses 4.3 $M_\oplus$ and 3.9 $M_\oplus$ with semi-major axes 0.36 and 0.46 au, respectively. To determine this the variation of the pulsar frequency had to be decomposed into the sum of two periodic variations of frequency, corresponding to the periods of the orbits, with different amplitudes, corresponding to each of the planet masses. Later, in 1994, a third planet was detected, closer to the pulsar with semi-major axis 0.19 au and mass 0.02 $M_\oplus$, less than twice the mass of the Moon. At first these observations were disputed because it was believed that a planet could not exist so close to a pulsar — after all a pulsar was the remnant of a supernova that would have destroyed everything in its vicinity. Nevertheless, the observations and their interpretation were sound and the planets might well have formed from the debris left behind by the supernova.

Several other discoveries have been made of planets around pulsars but the goal of detecting exoplanets associated with main-sequence stars was soon to follow.

13.2. Detecting Exoplanets Around Main-Sequence Stars

As previously mentioned, detecting an exoplanet has some of the characteristics of an astrometric binary in that we have two bodies in orbit around their combined centre of mass but only one of the bodies is visible — the star. However, the exoplanet case is much simpler in some ways, in that there is a large star:planet mass-ratio that enables some useful approximations to be made.

There are three quantities that can readily be found. The first is the mass of the main-sequence star, M_*, deduced from its spectral class. The second is the speed of the star in its orbit, v_*, found by Doppler-shift measurements, which have to be very precise, and the third is the period of the orbit, P, that comes from the periodicity of the Doppler-shift measurements. Assuming that the mass of the planet, M_P, is much less than that of the star we can find the radius of the planet's circular orbit, R, from

$$\omega^2 = \frac{4\pi^2}{P^2} = \frac{GM_*}{R^3}, \tag{13.2}$$

where ω is the angular speed in the orbits. We can now find the speed of the planet in its orbit, v_P, from

$$v_P = R\omega, \tag{13.3}$$

and hence the mass of the planet, M_P, from

$$\frac{M_P}{M_*} = \frac{v_*}{v_P}. \tag{13.4}$$

This analysis assumes that the line of sight is in the plane of the circular orbit. As explained in Section 11.6.2, if the line of sight makes an angle i with the line of sight then the deduced value of $v_\odot$ will be too small by a factor $\cos(i)$. Equation (13.2) will give the correct values for R and ω, and hence the value of v_P from (13.3) will also be correct. From equation (13.5) since the estimated value of v_* is too small then the estimated value of M_P is also too small by the same factor, $\cos(i)$.

Measurements are made from the Earth, moving relative to the Sun, so it is necessary to correct for its motion if the velocity of the star relative to the Sun $\nu_\odot$ is to be found. If the exoplanet orbit is circular then, relative to the Sun, a plot of the value of $v_\odot$ against time will be a sine curve (Figure 13.2). However, if the orbit is an ellipse then the plot would be a rather distorted version of a sine curve. (Figure 13.3) from which it is possible to get estimates of the exoplanet mass and the semi-major axis and eccentricity of its orbit. This profile is for a Jupiter-mass planet in orbit around a solar-mass star with a semi-major axis of 0.33 au and eccentricity 0.328.

The large ratio of the stellar mass to that of the planet means that the speed of the star is very small, a few metres per second, as are the Doppler shifts in spectral lines, requiring the use of very specialized optical spectrometers. Table 13.1 illustrates this by showing the contribution to the speed of the Sun due to Jupiter, Saturn and the

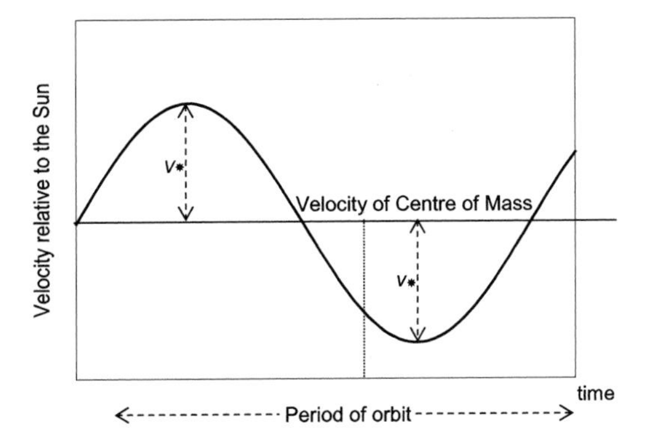

Figure 13.2. An ideal star velocity profile with a single exoplanet in a circular orbit.

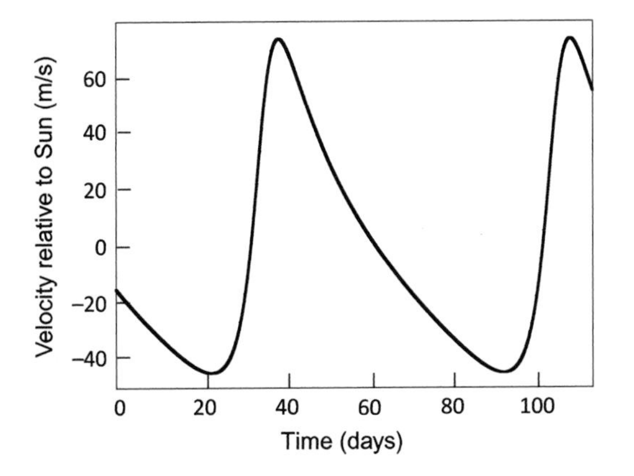

Figure 13.3. A typical velocity profile for a star with a planet in an eccentric orbit.

Table 13.1. The contributions to the speed of the Sun due to three planets.

Planet	Mass (M_ω)	Period (years)	Radius of orbit (km)	Speed (m s^{-1})
Jupiter	9.55×10^{-4}	11.86	7.78×10^{8}	12.4
Saturn	2.83×10^{-4}	29.46	1.43×10^{9}	2.7
Earth	3.00×10^{-6}	1.00	1.50×10^{8}	0.09

Earth. To have a larger orbital speed of the star requires a massive planet in a close orbit. Modern spectrometers are very accurate, giving speeds of less than $1\,\mathrm{m\,s^{-1}}$, although early measurements were made with instruments with an accuracy of somewhat greater than $10\,\mathrm{m\,s^{-1}}$.

The very first exoplanet detected accompanying a main-sequence star was that around the star 51 Pegasus by the French astronomers Mayor and Queloz (1995). The star has mass $1.06\,M_{\odot}$; the planet has a mass more than 0.468 times that of Jupiter and a circular orbit with radius 0.052 au. With an orbital period of just 4.23 days it was possible in a short time to observe several periods, which compensated for the low accuracy of their spectrometer — $13\,\mathrm{m\,s^{-1}}$.

Although Mayor and Queloz were the first to report the detection of an exoplanet, in 1996 Butler and Marcy later reported the existence of an exoplanet around the star 47 Uma after observations over a period of more than 8 years. Figure 13.4 shows the observations for a period of more than 10 years. The figure shows the best fit of a sine curve to the experimental measurements, which gave an orbital period of 1,094 days. From the mass of the star, $1.03\,M_{\odot}$, the orbital

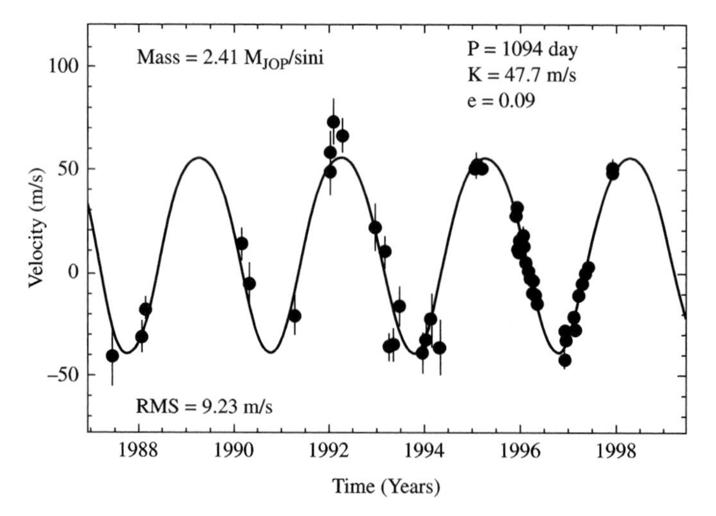

Figure 13.4. Velocity estimates for 47 Uma with the fitted velocity profile (Butler and Marcy, 1996).

radius of the planet is 2.09 au. The minimum mass of the planet is 2.62 times that of Jupiter.

Several hundred exoplanets have been discovered by astronomers examining individual stars that seemed likely to give a positive outcome but in 2000 the rate of discovery was greatly increased.

13.3. Transiting Exoplanets

If an exoplanet transits its star then more information can be found. Given that, as is usual, the distance of the exoplanet from the star is much greater than the radius of the star then the line of sight must make a small angle, i, to the orbital plane. Consequently $\cos(i)$ will be close to 1.0 and a good estimate is made for the actual mass of the planet rather than a lower bound that might be very different from the true mass. Another benefit of a transit is that it enables an estimate of the size of the planet to be made. Figure 13.5 shows the variation of the intensity of the light coming from the star as the transit progresses. While in transit, the planet blocks out some of the light coming from the star and the radius of the planet. r_p, is related to the radius of the star, r_* by

$$r_P^2 = r_*^2 \left(1 - \frac{b_2}{b_1}\right). \tag{13.5}$$

For main-sequence stars r_* is known from the spectral class.

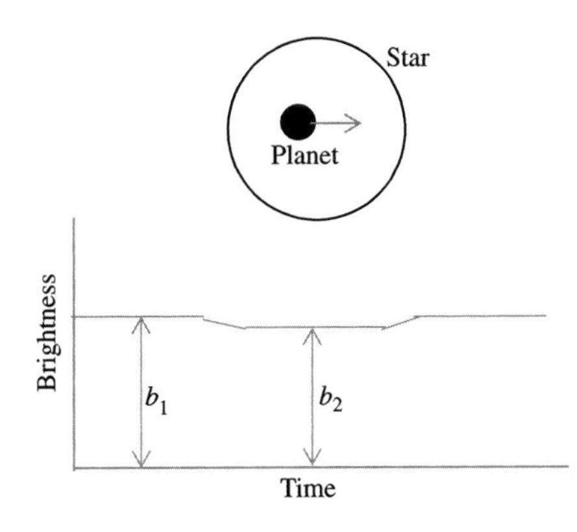

Figure 13.5. The variation of brightness as a planet transits a star.

In 2009, NASA launched the Kepler spacecraft which used a photometer to follow the variation of brightness of about 150,000 stars in a fixed patch of the sky. The data were then transmitted to Earth and analysed to find variations of brightness, as seen in Figure 13.5. Of particular interest in this mission was to detect Earth-like planets, which are easier to find this way than by detecting motions of the star using the Doppler effect. The Kepler mission has greatly increased the rate of discovery of exoplanets and by mid-2018 the total number found was approaching 4,000.

13.4. The Orbits of Exoplanets

The orbit of a planet is characterized by three quantities — semi-major axis, a, eccentricity, e and the inclination of the orbit, i, with respect to the equatorial plane of the star.

13.4.1. *Semi-Major Axis and Eccentricity*

The semi-major axis and eccentricity can be found by determining the motion of the star, as described in Section 13.2. Table 13.1 gives a selection of exoplanets showing typical ranges of masses, semi-major axes and eccentricities. There are some exoplanets that are so far from their stars that they can be directly imaged. Figure 13.6 shows the dust disk around the star Fomalhaut and the images of an exoplanet within the disk taken 2 years apart. It is difficult to find either the mass of the planet, which could be anywhere between a fraction of a Jupiter mass to three Jupiter masses, or the orbit, for which the best estimate is $(a, e) = (115\,\text{au}, 0.11)$. The largest known distance of a planet from a star is for a planet of mass $11\,M_J$ at 650 au from the star HD 106906 (Bailey *et al.*, 2014). At the other extreme there are exoplanets much closer to stars than Mercury's distance from the Sun, 0.39 au. Eccentricities also show a huge variation, from zero, corresponding to a circular orbit, to 0.9349 for the planet orbiting HD 80606. At its closest this planet is distant 0.03 au from the star and at its furthest 0.9 au. It has been estimated from observations that when it swings past the star at closest approach its surface temperature goes from 800 K to 1 500 K in the space of 6 h.

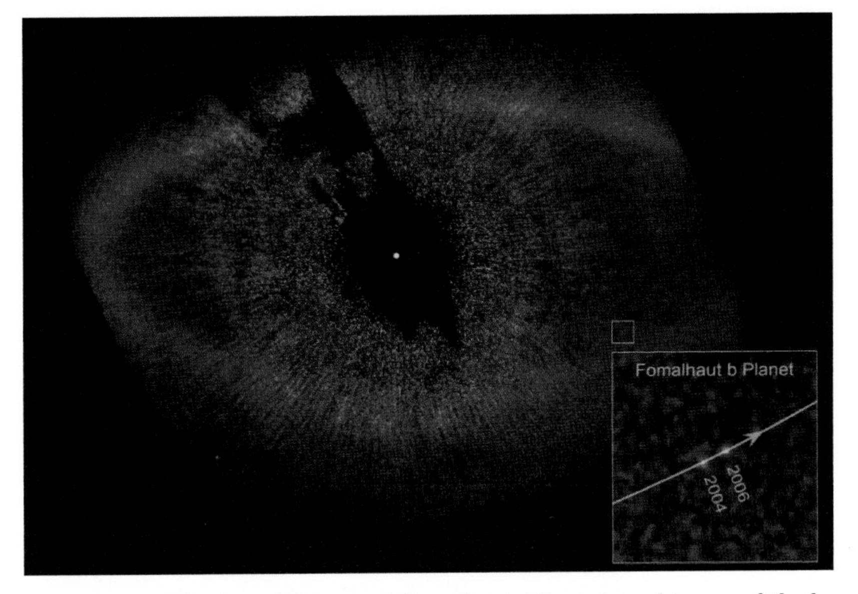

Figure 13.6. The dust disk around Fomalhaut. The enlarged image of the box shows the planet in 2004 and 2006 (NASA/ESA).

There are several other interesting features of the exoplanets listed in Table 13.2. There is a family of five planets associated with the star 55 Cancri and there are three planets in the family of GJ581. The masses of the planets vary from 18.4 M_J, which is actually a brown-dwarf mass, to 0.0159 M_J — about 5 times the mass of the Earth.

13.4.2. *Inclination (Spin-Orbit Misalignment)*

In 2009 the transiting exoplanet WASP-17b was found to be in a retrograde orbit around its star (Anderson *et al.*, 2010). Several other retrograde orbits have been found since from Kepler-mission observations using the Rossiter-McLaughlin effect (Rossiter 1924; McLaughlin, 1924). A spinning star will generally have a component of the total Doppler shift that depends on the distance from the spin axis. A boundary of the observed star that is moving away from the observer relative to the centre of the star will have a relative red shift and, conversely, the boundary on the opposite side will show

Table 13.2. The characteristics of some exoplanets.

Star	Minimum planet mass (M_J)	Semi-major axis (au)	Eccentricity
HD41004B	18.4	0.0177	0.081
GJ436	0.0692	0.0278	0.159
55Cancri	0.816	0.114	0.0159
	0.165	0.238	0.053
	3.84	5.84	0.063
	0.0235	0.0377	0.264
	0.141	0.775	0.002
HD72659	3.30	4.77	0.269
HD80606	4.31	0.468	0.9349
GJ581	0.0490	0.0406	0.000
	0.0159	0.0730	0.000
	0.0263	0.253	0.000

a relative blue-shift. The situations for inclinations of 0° (prograde) and 180° (retrograde) are shown in Figure 13.7; looking at the star from the top it is spinning clockwise. In the absence of a planet the light seen would include a range of wavelengths with a peak intensity somewhere between the peak wavelengths at the two boundaries. Now, if the planet moves over the disk of the star in the direction from B to A its orbit will also be clockwise. When it first enters the disk it will block out some blue-shifted wavelengths and the peak wavelength will move towards the red end of the spectrum. Just before it leaves the disk it covers part of the red-shifted region and the peak wavelength is moved towards the blue end of the spectrum. This pattern, where the peak goes from red-shifted towards blue-shifted relative to the no-transit state, is indicative of low inclination — 0° in our simple illustration. Conversely, if the motion of the planet is in the direction from A to B the inclination is 180°, i.e. retrograde, and the peak wavelength will go from blue-shifted towards red-shifted.

Figure 13.7 is just a simple illustration of the principle of the Rossiter–McLaughlin effect where the spin axis of the star is perpendicular to the line of sight and the planetary orbit is in the

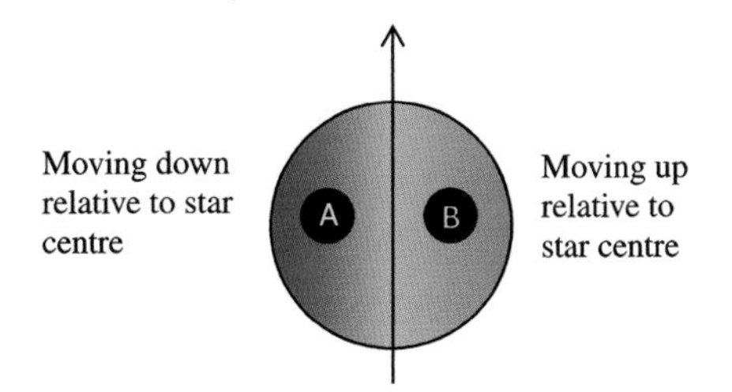

Figure 13.7. An illustration of the Rossiter–McLaughlin effect.

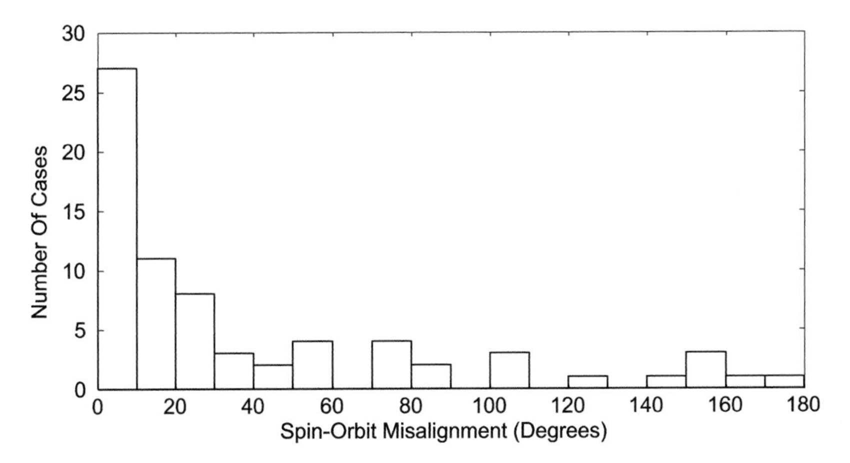

Figure 13.8. Spin-orbit misalignments (inclinations) for exoplanets (Woolfson, 2013a).

plane of the stellar equator. However, different forms of the progress of colour-shifting of the peak with time enable other inclinations to be determined.

Another method for measuring inclinations depends on the measurement of blips on the transit intensity curve due to star spots (Nutzman, Fabrycky and Fortney, 2011). The Figure 13.8 histogram is derived from spin-orbit misalignment (SOM) — i.e. inclination — observations given by Rene Heller in 2013. Retrograde orbits (SOM > 90°) are common, although prograde orbits, especially with small SOMs, are predominant, including planets of the Solar System with average SOM 7°.

13.5. Other Observations

Planets have been observed not only around isolated stars but also around one or both of the members of binary systems. The stars of the system, γ-Cephei have masses 1.40 $M_\odot$ and 0.40 $M_\odot$ with orbital parameters $(a, e) = (20.2\,\text{au}, 0.41)$. An exoplanet orbits the more massive star with orbital parameters $(a, e) = (2.04\ \text{au}, 0.115)$ (Observatoire de Paris, 2004).

Another observation, based on the totality of exoplanet discoveries, is that a large proportion of stars have one or more planetary companions. Estimates of the proportion of stars with planets vary with time and tend to increase; a more recent estimate is 0.34 (Borucki *et al.*, 2011), although there are some higher estimates.

13.6. Other Features Associated with Planets

The only planetary system of which we have detailed knowledge is the Solar System and it might be expected that some properties of the Solar System would also be part of exoplanet systems. For example, all the major Solar System planets have extensive families of satellites, so it seems very likely that satellite formation is a natural concomitant of planet formation. Again, the Solar System contains four major planets and they are connected in pairs by having commensurate orbital periods, close to the ratio of small integers, e.g.

$$\frac{\text{Orbital period of Saturn}}{\text{Orbital period of Jupiter}} = \frac{29.46\,\text{years}}{11.86\,\text{years}} = 2.48 \approx \frac{5}{2},$$

$$\text{and} \quad \frac{\text{Orbital period of Neptune}}{\text{Orbital period of Uranus}} = \frac{164.8\,\text{years}}{84.02\,\text{years}} = 1.96 \approx \frac{2}{1}.$$

Finally we note that the Solar System contains both terrestrial-type planets and also major planets, the latter being of two distinct types — gas giants and ice giants. Observations have indicated the presence of rocky terrestrial-type and major-type exoplanets but the division into two kinds of major planet cannot be reliably determined with present technology.

13.7. Requirements for a Plausible Theory of Planet Formation

First it is necessary to consider how theories should be judged, especially in fields where it is not possible to do experiments and where information is mostly obtained from observations. The first proposition is that no theory can be claimed to be correct, even if it satisfies all the constraints imposed by observations and stays within the bounds of accepted scientific principles. On the other hand, a theory can be judged as wrong if it is in conflict with either observations or with well-established scientific principles, such as the standard conservation laws of physics. If there is more than one theory that satisfies the requirements for plausibility then each one must be given some measure of acceptance until it is in irreconcilable conflict with either some well-tested observation or new application of theory.

In view of the profusion of exoplanets and the range of types of planets observed a *sine qua non* for any theory is that it must involve processes that will occur frequently and will give both terrestrial-type and major planets. When the only known planetary system was the Solar System then theories involving highly unlikely events could not be discounted — and many past theories were in that category.

The whole range of orbital parameters of exoplanets must also be explained. The semi-major axes vary from fractions of an astronomical unit to several hundred astronomical units and eccentricities from zero to close to unity. Perhaps the most intriguing orbital property is when the spin-orbit misalignment is greater than 90° corresponding to a retrograde orbit. All the SOMs of solar-system planets are small and we are conditioned to think that this is normal and that retrograde orbits are abnormal. Retrograde orbits are clearly a minority — but a substantial minority.

Unless Solar System formation of planets is different from the common way of forming planets then satellite formation should be explained as a natural concomitant of planet formation. Similarly, assuming that the Solar System is a typical planetary system then some explanation of orbital commensurabilities, which also occur with some sets of satellite orbits, would be a welcome addition to any theory.

The Solar System contains many small bodies that may be the result of particular evolutionary features and hence not typical of all planetary systems. This includes the Kuiper Belt a region with inner boundary just beyond the orbit of Neptune, which contains many small comet-like bodies and also a small number of *dwarf planets*, satellite-size bodies. At a distance of tens of thousands of astronomical units there is a reservoir of cometary bodies in what is known as the *Oort Cloud*; the origin of which, whether it originated with the Solar System or is of external origin, has been hotly debated. Finally there is the Asteroid Belt, tens of thousands of small rocky or iron bodies mostly situated between Mars and Jupiter. Any explanation of the Solar System features that depended on the basic mechanism of planet formation being proposed would be a bonus for any theory but not necessary to give plausibility.

Now we are going to describe two extant theories of planet formation that satisfy many, if not all, of the conditions for a plausible theory.

Problems 13

13.1 A star of mass $1.2\,M_\odot$ is found from Doppler-shift measurements to be moving on a circular orbit with speed $9.0\,\mathrm{ms}^{-1}$ with a period of 4.62 years. What is the minimum mass of the accompanying planet?

13.2 A star of mass $1.3\,M_\odot$ and radius $1.2\,R_\odot$ is transited by a planet with a period of 2.83 years. The orbital speed of the star is found to be $11\,\mathrm{ms}^{-1}$. During the transit the brightness of the star reduces to 0.972 of its non-transit value. Determine the mass and radius of the planet.

13.3 The following table gives the periods of some of the major satellites of Saturn. Find e period commensurabilities between pairs of satellites.

Satellite	**Period (Days)**
Mimas	0.9424
Enceladus	1.3702
Tethys	1.8878
Dione	2.7369
Rhea	4.5175
Titan	15.9454
Hyperion	21.2766

Chapter 14

The Nebula Theory

14.1. The Laplace Nebula Theory

The first theory of the origin of the Solar System based on scientific principles was proposed at the end of the 18th century by the eminent French mathematician and astronomer, Pierre-Simon Laplace (1749–1827; Figure 14.1), in his book *Exposition du Système du Monde* (Laplace, 1796). The mechanism he proposed is shown in Figure 14.2. The model began with a large, slowly-rotating nebula — a sphere of dusty gas that was collapsing under the influence of self-gravitational forces (Figure 14.2(a)). As the nebula collapsed so, to conserve angular momentum, the rate of spin increased, causing the nebula to take on the shape of an oblate spheroid (Figure 14.2(b)). Eventually the spin rate reached a value such that equatorial material was in free orbit around the central mass and the nebula took on a lenticular shape — like a biconvex lens with a sharp edge (Figure 14.2(c)). Thereafter, as the nebula continued to collapse it left behind orbiting material in the equatorial plane. At this stage Laplace made the assumption that the loss of material at the boundary would be spasmodic so that material left behind formed a series of concentric rings (Figure 14.2(d)). In the next stage of the process, material in each ring clumped together under gravitational forces with several clumps in each ring (Figure 14.2(e)) and since the clumps would have slightly different angular velocities they eventually amalgamated to give a single condensation that would collapse to form a planet

Figure 14.1. Pierre-Simon Laplace.

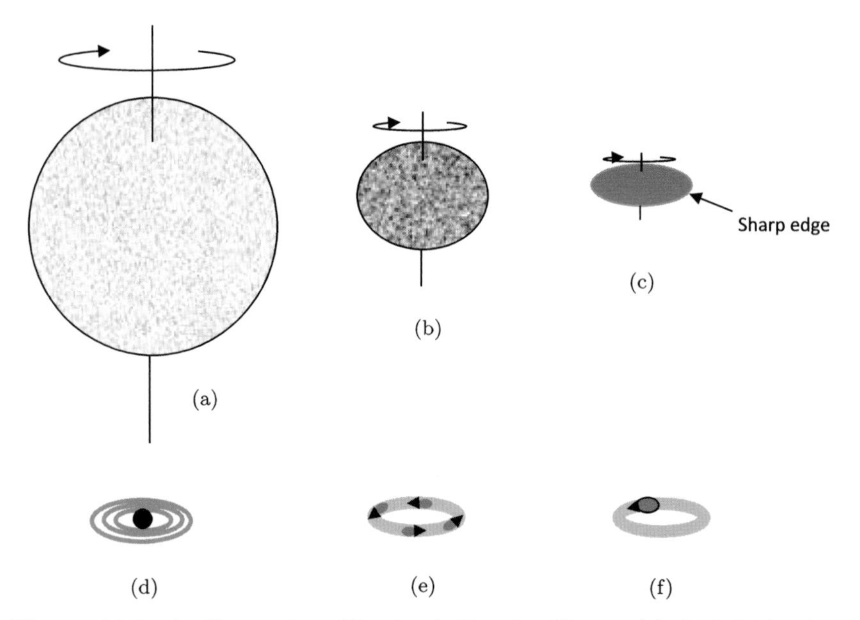

Figure 14.2. An illustration of Laplace's Nebular Theory. (a) the initial sphere of gas and dust. (b) forming an oblate spheroid (c) a lenticular shape with a sharp edge (d) the formation of gas rings (e) condensations forming within a ring (f) coalescence to form a planetary blob.

(Figure 14.2(f)). The central condensing mass of material formed the Sun. It was a *monistic theory*, producing the Sun and planets from the same body of material.

This theory became the standard model of solar-system formation, with wide acceptance by the scientific community. It produced both the Sun and planets from a very plausible starting point and explained the major features of the Solar System, with planets orbiting in the same sense as the spin of the Sun in nearly coplanar orbits.

By the middle of the 19th century serious doubts were being expressed about the plausibility of the theory, the main problem being concerned with the distribution of angular momentum in the Solar System. The Sun, with 99.86% of its total mass spins slowly on its axis and contains only 0.5% of its angular momentum, the remainder being contained in the planetary orbits. It was difficult to envisage how Laplace's nebula could evolve in such a way that 0.14% of the mass ended up with 99.5% of the angular momentum.

There are other ways of expressing this angular-momentum problem. If at the time that the Neptune ring formed the whole nebula had the angular velocity of Neptune in its orbit then the total angular momentum of the nebular would be hundreds of times that of the actual Solar System. Alternatively, accepting the present angular momentum of the Solar System, then at the time the nebula had the radius of Neptune's orbit its angular velocity would have been far too small to have left behind a Neptune ring.

There were attempts to rescue the theory. The French astronomer Edouard Roche (1820–1883), proposed extreme distributions of material in the initial nebula to explain the partitioning of mass and angular momentum but this led to too little mass at the outside of the system to form the outer planets. By the end of the 19th century Laplace's nebular theory was abandoned and other ideas were being considered.

14.2. Revisiting Nebula Ideas and the Angular Momentum Problem

Meteorites are, for the most part, fragments of asteroids produced by collisions of those bodies and meteoriticists — workers in the meteorite field — look for clues in meteorites about the early history of the Solar System, or even about its origin. By the early 1960s, it was clear that at some stage in the early Solar System, perhaps even at its origin, the Solar System had been engulfed in a hot silicate-plus-iron vapour. There were several indications of this, there are *chondritic meteorites*, stony meteorites that contain *chondrules*, small millimetre-size silicate spheres that are frozen droplets that condensed out of a hot vapour. On the surfaces of some cavities in meteorites there are silicate crystals that have been deposited directly from a vapour and, finally, meteorites show evidence of condensation sequences where there has been a direct transition from the vapour to the solid phase based on a sequence of decreasing condensation temperatures (Grossman, 1972).

The undoubted presence of a hot vapour in the early Solar System gave rise to the idea that the Solar System formed from a hot nebula and there was a resurgence of interest in nebula-theory ideas. Although Laplace's theory had failed it was felt that new kinds of process, that Laplace could not have known about, would provide the answer to the problems that caused the original model to fail. The first idea was that magnetic fields could transfer angular momentum from the centre of the nebula to its outer regions and this could solve the problem of the partitioning of mass and angular momentum. Then it was assumed that the disk forming around the central condensation would become gravitationally unstable and the blobs so formed would condense into planets. However, the idea of forming planets in this way from a high-temperature disk was soon found to be untenable and a very prominent worker in the new nebula theory, the Canadian astrophysicist A. G. W. Cameron (1925–2005) wrote in 1978:

> 'At no time, anywhere in the solar nebula, anywhere outwards from the orbit of Mercury, is the temperature in the unperturbed solar nebula ever high enough to evaporate completely the solid materials contained in interstellar grains.'

While this seemed to undermine the very *raison d'être* for considering nebula ideas again, by this time a great deal of thought and effort had gone into developing a new nebula theory and the work continued. The new theory that was being developed did not depend on a high-temperature nebula.

What is required to solve the problem of partitioning mass and angular momentum in the solar-system context is the transfer of angular momentum from inner to outer material as the nebula collapses. Several ways in which this might happen have been suggested.

14.2.1. *Angular Momentum Transfer by a Magnetic Field*

One possible mechanism was suggested by the British astrophysicist Fred Hoyle (1915–2001) in 1960, just as the first thoughts about a new nebula theory were beginning to be formulated. He assumed that, in the evolving nebula, a gap had opened up between inner and outer material and that the gaseous material on both sides of the gap was at a sufficiently high temperature for it to become ionized, and hence conducting. If the central core of the nebula generated a magnetic field then field lines would connect the inner and outer material and become frozen into the conducting material. These field lines would behave like elastic strings that were firmly attached at both ends (Figure 14.3). As the central material collapsed its spin rate would tend to increase thus stretching the field lines; this would create forces pulling backwards on inner material, so slowing down its spin and causing it to lose angular momentum, and forward on outer material, producing an increase in its angular momentum. This is a transfer of angular momentum from inner to outer material — just what is required. There are problems with this process. It would require a large magnetic field about one thousand times that of the Sun — to be effective. For a small field, if the magnetic field lines stayed intact, they would form a spiral around the core but what would happen in practice is that the energy associated with the stretched field lines would reduce by the field lines breaking and reforming to give shorter field lines. Again it has to be shown that temperatures in an early nebula would be high enough to give the required ionization of disk material. Regardless of this, a new

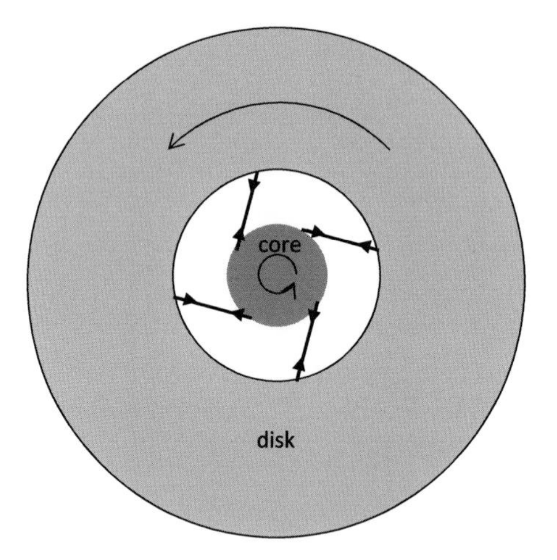

Figure 14.3. Hoyle's model of flux lines linking a collapsing core to a disk.

concept had been introduced, that of using magnetic fields to transfer angular momentum outwards, and it was felt that some other process involving magnetic fields might be forthcoming that could do what is required.

14.2.2. *The Armitage and Clarke Mechanism*

Several more ideas, differing in their details and complexity, have been suggested to use magnetic fields to transfer angular momentum from inner to outer material. A typical one, worked out in some detail, was described by Armitage and Clarke (1996).

The core of the early star generates a magnetic field, about one thousand times that of the present Sun. There is a disk in orbit around the core with material in the inner part of the disk orbiting at a rate faster than that of the central core. However, the disk material further out is orbiting at a lower rate, following a Kepler-rate-like reduction in angular speed with distance. The magnetic linkage is assumed to occur between the core and all parts of the disk, as illustrated in Figure 14.4 in edge view. The magnetic flux lines linking the core to the more slowly rotating outer material transfer angular momentum outwards, as Hoyle had described, so that material moves outwards. The very strong magnetic field causes

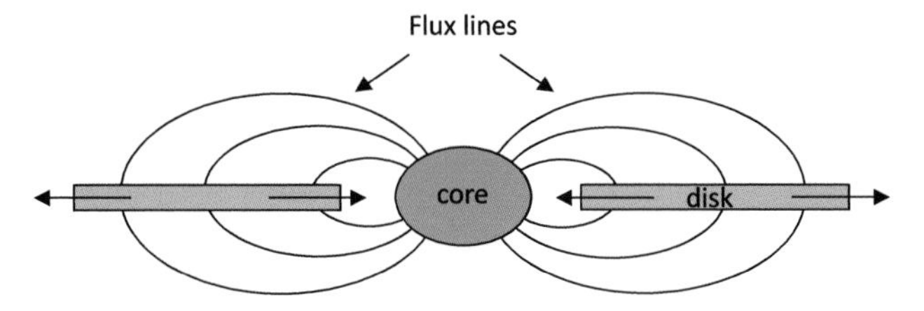

Figure 14.4. Inner disk material moves inwards to join the core while outer disk material moves outwards.

the field lines to behave like very powerful elastic strings that have to be stretched very little to give the required forces and therefore they do not break and reconnect. The more rapidly orbiting material in the inner part of the disk is also connected to the core by flux lines but these give a different form of behaviour. The inner part of the disk is now assumed to become rather turbulent and material from this part of the disk flows along the flux lines to join the central core.

For this model to work requires several conditions to be satisfied. There must be a magnetic field of sufficient strength and of a form that would enable a strong linkage to occur. Another requirement is that the disk material should be hot enough to be conducting at some considerable distance from the star. That should not be a problem at the inner edge of the disk, which is close to the star and fully exposed to its radiation. Another problem is that, for a star to form, angular momentum transfer must occur, but for angular momentum to be transferred a strong magnetic field is needed, requiring the existence of a star. In short, the starting point of the model needs justification.

14.2.3. *A Mechanical Process for Transferring Angular Momentum*

Although most ideas about transferring angular momentum involved the application of magnetic fields in some way, there was an important contribution that depended purely on mechanical forces. In 1974 the British astrophysicists, Donald Lynden-Bell (1935–2018) and James Pringle (b. 1949), analysed what would happen if in an isolated core-plus-disk system lost energy in some way — perhaps due to internal viscous forces within the disk or by material falling on the disk from outside, both of which would convert mechanical energy into heat energy that would then be radiated away. It is straightforward to show that in an isolated spinning system that loses energy but conserves angular momentum the net result is that inner material moves inwards and outer material moves outwards. Consider two bodies in circular orbits, with radii r_1 and r_2 around a central body of much greater mass. The energy, E, and angular momentum, J, of the system are of the form

$$E = -C\left(\frac{1}{r_1} + \frac{1}{r_2}\right), \tag{14.1}$$

and

$$J = K\left(\sqrt{r_1} + \sqrt{r_2}\right), \tag{14.2}$$

where C and K are two positive constants. For small changes in r_1 and r_2 the changes in E and J are:

$$\delta E = C\left(\frac{1}{r_1^2}\delta r_1 + \frac{1}{r_2^2}\delta r_2\right), \tag{14.3}$$

and

$$\delta J = \frac{1}{2}K\left(\frac{1}{\sqrt{r_1}}\delta r_1 + \frac{1}{\sqrt{r_2}}\delta r_2\right). \tag{14.4}$$

If angular momentum remains constant then, from (14.4)

$$\delta r_1 = -\sqrt{\frac{r_1}{r_2}}\delta r_2, \tag{14.5}$$

and substituting for δr_1 in equation (14.3) gives

$$\delta E = \frac{C}{\sqrt{r_2}}\left(\frac{1}{r_2^{3/2}} - \frac{1}{r_1^{3/2}}\right)\delta r_2. \tag{14.6}$$

If δE is negative and if $r_2 < r_1$ then δr_2 must be negative, i.e. the inner body moves inwards and hence, from equation (14.5) the outer body moves outwards. This outward transfer of angular momentum is true, not just for two bodies but also for a general distribution of material — in particular a core with a surrounding disk. With inner material moving inwards and outer material outwards a gap might be created between core and disk across which a magnetic field may act further to transfer angular momentum.

14.2.4. *Angular Momentum Distribution in a Newly Formed Star*

Section 4.8 gave a model for star formation by the collision of turbulent elements within a collapsing star-forming cloud. In some circumstances, after rapid cooling, the high density region produced by the collision would have greater than the Jeans critical mass and continue to collapse. The angular momentum relative to the centre of mass of the region could be anything from zero up to a very large value. If very high then the region would fly apart and no star would form. For low values the region would collapse to the point of being a YSO while still being rotationally stable for intermediate angular momenta the region could bifurcate, giving a binary system, each member of which would form a stable star. In the case of a single star the operation of the Lyndon-Bell and Pringle mechanism would transfer material outwards to form a disk. Another contribution to a disk could be peripheral material attaching itself to the collapsing core. The Sun, in common with stars of around its mass and less, spins slowly and it is very unlikely that a newly-formed star would be spinning that slowly and the partitioning of mass and angular momentum in the YSO and disk would not be anything like that found for the Sun and solar-system planets. However, the star would still be rotationally stable and there is another mechanism, again dependent on a magnetic field, that slows down the initial spin rate of the star by a considerable amount.

14.2.5. *Magnetic Braking of Stellar Spin*

Although the mechanism to be described here is applicable to all stars we will restrict our attention to the Sun, the spin of which is slow, even for a G2 star. A mechanism for slowing down its original spin involves coupling of ionized solar-wind material moving out of the Sun with the solar magnetic field (Cole and Woolfson, 2013).

A charged particle leaving the Sun, couples to the Sun's magnetic field by moving on a helical path around a field line. In the vicinity of the Sun, where the field is strong, it is strongly coupled but, as the field weakens further out, the radius of the helical path increases and finally the charged particle detaches itself from the field line. The magnetic field rotates with the Sun, so the escaping material will co-rotate with the Sun while it is moving outwards and still coupled. Thus the escaping solar wind material gains angular momentum, which is removed from the Sun.

Now we examine this mechanism in more detail and first consider how the charged particles initially couple to field lines and then later become decoupled. This depends on the relative strengths of the magnetic pressure (energy density) given by

$$P_B = \frac{B^2}{2\mu_0}, \tag{14.7}$$

in which B is the strength of the field and μ_0 the permeability of free space, and the total gas pressure

$$P_g = nkT + nmv^2, \tag{14.8}$$

in which T is the temperature and n the number density of particles of mean mass m and bulk flow velocity v. The first term in (14.8) is the normal gas pressure and the second term is the dynamic pressure due to the bulk flow — like the pressure on a sail due to the wind. If the magnetic pressure is greater, then the motion of the charged particles is controlled by the field and it remains coupled to a field line. Equality of the pressures gives the situation in which the particles decouple from the field.

The form of the Sun's magnetic field is quite complex because of the interplay of the field with the solar wind. Far from the Sun, where the field is weak, the solar wind is virtually unconstrained by the magnetic field, which becomes frozen into the solar wind. The solar magnetic field is very complex, especially near its very dynamic surface where there are local fields generated, but at some distance

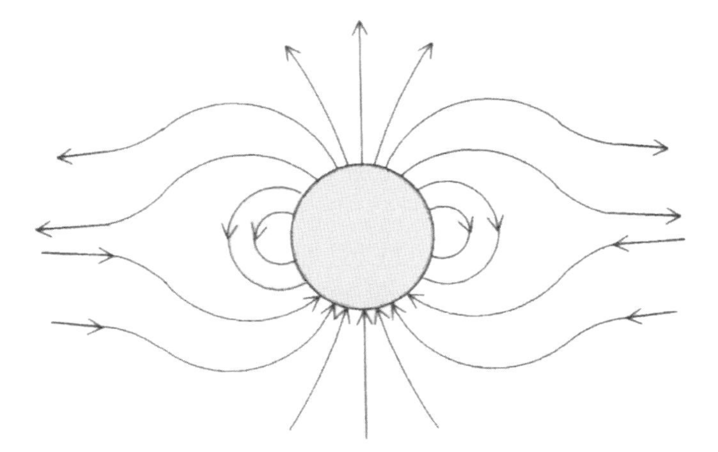

Figure 14.5. Magnetic field lines round a star with a strong stellar wind of ionized particles.

from the surface the field can be approximated as

$$B = D_{\odot}\left\{\frac{1}{r^3} + \frac{1}{(30R_{\odot})^2 r}\right\}, \tag{14.9}$$

in which $D_{\odot}$ is the magnetic dipole moment, $R_{\odot}$ the radius of the Sun and r the distance from the centre of the Sun. (Freeman, 1978). The corresponding arrangement of field lines is schematically illustrated in Figure 14.5.

We can now work out what the rate of loss of angular momentum of the Sun may have been in the past. For protons as the solar-wind particles and with a solar-wind speed of $500\,\mathrm{km\,s^{-1}}$ the second term on the right-hand side of (14.8) will dominate unless T is greater than about $10^7\,\mathrm{K}$. It also turns out that, up to distances where decoupling takes place, the field is effectively of dipole form, i.e. one can ignore the second term in the bracket in equation (14.9). Equating the magnetic and gas pressures in the equatorial plane with these simplifications

$$\frac{D_{\odot}^2}{2\mu_0 r^6} = nmv^2. \tag{14.10}$$

The local density of the ionized material is nm and in terms of the rate of mass loss from the Sun, dM/dt, assuming that all lost material is ionized and is emitted equally in all directions,

$$nm = \frac{dM/dt}{4\pi r^2 v}. \tag{14.11}$$

Equating values for nm in (14.10) and (14.11) we find that co-rotation of ionized material will persist out to a distance in the equatorial plane

$$r_c = \left(\frac{2\pi D_\odot^2}{\mu_0 v \frac{dM}{dt}} \right)^{\frac{1}{4}}. \tag{14.12}$$

Present values for the Sun are: $dM/dt = 2 \times 10^9\,\mathrm{kg\,s^{-1}}$, $D_\odot = 8 \times 10^{22}\,\mathrm{T\,m^3}$ and $v = 5 \times 10^5\,\mathrm{m\,s^{-1}}$, give $r_c = 3.4\,R_\odot$. If all the mass were expelled in the equatorial plane of the Sun then the lost mass would take from the Sun $(3.4)^2$ times the angular momentum it had when it was part of the Sun. However, mass is lost from all the Sun's surface and moves off in all directions. How to allow for this in estimating the removal of angular momentum from the Sun is fairly complicated. On the one hand, for a given distance of decoupling, the distance from the spin axis of the Sun is reduced for non-equatorial material. On the other hand, for a given distance from the Sun the field is stronger further from the equatorial plane (twice as great for a given distance along the dipole axis) so the decoupling distance increases. What we do to allow for this is to take the value of r_c^2 as β times the value of r_c^2 indicated by (14.12) and $\beta = 0.5$ will give a reasonable ballpark estimate of the slowing-down process.

At the present solar-wind rate the total mass loss of the Sun over its lifetime would have been

$$\delta M = \frac{dM}{dt} t_\odot, \tag{14.13}$$

where $t_\odot$ is the age of the Sun, which we take as 4.5×10^9 years or 1.42×10^{17} s. The total loss of mass would have been 2.8×10^{26} kg, or 1.4×10^{-4} of the Sun's mass.

The early Sun, especially as a YSO, would have had both a greater rate of mass loss and a larger magnetic field. The magnetic field could have been several thousand times as strong the rate of loss of mass was far higher than at present. The usual assumption is that the early Sun went through a T-Tauri stage when the loss of mass was of order $10^{-7} M_\odot\,\text{year}^{-1}$ ($\sim 6 \times 10^{15}\,\text{kg}\,\text{s}^{-1}$), sustained for a period of 10^6 years. Such a rate of loss is at the upper end of expectations and we shall also consider smaller rates.

The magnetic braking effect is able to reduce the original spin rate of the Sun to a small fraction of its original value with rates of loss of the Sun's mass and magnetic fields well within the bounds of speculation. We assume that the moment of inertia of the Sun is always of the form $\alpha M R_\odot^2$, where M is the changing mass, but that the radius is constant. The rate of loss of angular momentum of the Sun is then given by

$$\frac{dH}{dt} = \frac{dM}{dt}(\beta r_c^2 - R_\odot^2)\Omega, \tag{14.14}$$

where Ω is the spin angular speed. The expression for angular momentum is

$$H = \alpha M R_\odot^2 \Omega,$$

which gives

$$\frac{dH}{dt} = \alpha R_\odot^2 \Omega \frac{dM}{dt} + \alpha R_\odot^2 M \frac{d\Omega}{dt}. \tag{14.15}$$

From (14.14) and (14.15)

$$\frac{d\Omega}{dt} = \left(\frac{\beta r_c^2}{\alpha R_\odot^2} - \frac{1+\alpha}{\alpha}\right)\frac{\Omega}{M}\frac{dM}{dt}, \tag{14.16}$$

which gives

$$\frac{d\Omega}{dM} = C\frac{\Omega}{M}, \tag{14.17}$$

with $C = \frac{\beta r_c^2}{\alpha R_\odot^2} - \frac{1+\alpha}{\alpha}$.

Table 14.1. The fraction of the initial angular momentum remaining after 5×10^5 years for combinations of magnetic dipole moment and rates of mass loss.

$dM/dt(M_\odot$ year$^{-1})$	$D/D_\odot$				
	1,000	2,000	3,000	4,000	5,000
10^{-9}	0.749	0.554	0.411	0.305	0.226
10^{-8}	0.427	0.165	0.064	0.025	0.009
10^{-7}	0.124	0.006	3×10^{-4}	10^{-5}	6×10^{-7}

Integrating (14.17) gives

$$\Omega = \Omega_0 \left(\frac{M}{M_0}\right)^C, \tag{14.18}$$

where M and M_0 are the final and initial masses and Ω and Ω_0 are the final and initial spin angular speeds. For a constant rate of mass loss M is related to M_0 by

$$M = M_0 - \frac{dM}{dt}t,$$

where t is the duration of the mass loss.

The effect of different combinations of rate of loss of mass of 10^{-9}, 10^{-8}, and $10^{-7}M_\odot$ per year with a magnetic dipole moment from 1,000 to 5,000 times the present value, is shown in Table 14.1 with α (angular momentum factor for the Sun) $= 0.055$. The duration of the mass loss is taken as 5×10^5 years with the radius remaining constant at its present value. This last assumption is certainly not true, since a star in a T-Tauri state is evolving towards the main sequence with both a reducing radius and a changing moment-of-inertia factor, but the general thrust of the way the spin reduces is illustrated by the results in the table. It is clear that combinations of rate of loss and dipole moment towards the upper ends of the ranges considered are capable of giving a very large reduction of angular momentum.

Starting with a spherical nebula, the process of turning it into a star plus disk, as Laplace sought to do, presents a number of problems that have not yet been convincingly solved by the application of magnetic linkage between inner and outer material — although such a solution cannot be ruled out A more straightforward possible solution is to postulate star production by the collision of streams of gas in a turbulent star-forming region that can give either an isolated star or a binary system in which the stellar spin rates can be up to two orders of magnitude greater than at present, with the star still being rotationally stable. Subsequently the removal of angular momentum by a stellar wind coupled to the stellar magnetic field can reduce the stellar spin rate to a small fraction of its original value.

The presence of a disk around the star can then be due to a combination of the retention of the original compressed region of some peripheral material, or the capture of material and the action of the Lynden-Bell and Pringle mechanism.

14.3. The Formation of Planets

When the return to nebula ideas began, the new theory was called the Solar Nebula Theory, because the Solar System was the only planetary system known and it was not certain that any other existed. There was the assumption that young stars had accompanying disks but that assumption was not confirmed until the mid-1970s when disks were detected around young stars (Beckwith and Sargent, 1996). The detection came from a careful examination of the spectra of young stars, which mostly resembled what is seen in Figure A.4 except that there were small bumps in the far infrared regions. This corresponded to output from a very cold source but, since cold radiators are very inefficient and since there was a bump that could be detected, it was inferred that the area of the source had to be large. The interpretation was that these bumps indicated the existence of a cold disk accompanying the star. Later, disks around stars were imaged and were estimated to have masses between one-hundredth to one-tenth of a solar mass. However, it was also deduced that these disks were comparatively short-lived with a typical lifetime of 3 million years and a maximum lifetime of about 10 million years.

When exoplanets were discovered the name of the theory changed to the Nebula Theory (NT) since it was now necessary to explain the existence and properties of all the planets discovered and not just those of the Solar System.

In fact the NT was not the first 20th century theory to postulate that planets were formed in cold material accompanying a star. In 1944, the Russian planetary scientist Otto Schmidt (1891–1956) suggested that the Sun had once passed through a cool dark cloud and on emerging had acquired an envelope of dusty gas but he did not develop in detail how this material would produce planets.

14.3.1. *Converting a Dusty Disk into Planets*

It was a colleague of Otto Schmidt, Victor Safronov (1917–1999), who, in 1968, was the main architect of the theory of producing planets from a dusty disk that is now generally accepted. A translation of his work into English is available in Safronov (1972). The process, as originally described, has four main stages:

(i) Under gravitational forces, the dust settles towards the mean plane of the disk.
(ii) Gravitational instability of the dust carpet produces solid bodies of kilometre to 100 km dimensions called *planetesimals.*
(iii) Planetesimals aggregate to form terrestrial-type planets or the cores of major planets.
(iv) The cores of major planets capture gas from the disk to form extensive atmospheres.

We will now discuss these stages individually in more detail.

14.3.2. *Forming Planetesimals*

Although the material around the star is called a ‘disk’ it cannot be in the form of a coin with uniform thickness, for such a configuration would be unstable. The areal density, i.e. density per unit area would fall off with distance from the star but the thickness of the disk would increase with distance with the density a maximum in the mean plane of the disk and falling off on both sides perpendicular to the disk.

The original disk material is very diffuse; taking a mass of one-tenth of a solar mass with a radius of 30 au the mean areal density is just over 3,100 $\text{kg}\,\text{m}^{-2}$. It would be much smaller in the outer regions of the disk and would be spread perpendicular to the disk over a greater distance — of order 10 au at the outer edge. Producing a dust disk concentrates the material for forming planetesimals.

An argument that has been raised in favour of planetesimal formation is that they are of asteroid size or greater and that planetesimals still exist in the form of asteroids and comets. Two processes have been suggested for planetesimal formation. The first, and probably the most favoured, is through the gravitational instability of the dust carpet. This will tend to produce clumping of dust but disruptive solar tidal forces will oppose this tendency. The theory for dust-carpet gravitational instability was given by Goldreich and Ward (1973). For a planetesimal clump to form at distance R from a star it must be able to withstand disruption due to the tidal forces produced by the star. This is when

$$\rho_P \geq \frac{3M_*}{2\pi R^3}, \tag{14.19}$$

in which ρ_P is the mean density due to the local distribution of planetesimals and M_* the stellar mass. When the dust carpet in a particular location has a thickness, h, such that its density reaches the critical level, then gravitational instability will occur and the carpet will break up. If the areal density of the dust component is ρ_{ad}, then that thickness is given by

$$h = \frac{\rho_{\text{ad}}}{\rho_P}. \tag{14.20}$$

Safronov (1972) estimated that the area of a condensation will be about $60\,h^2$ so the total volume of a planetesimal condensation is about $60\,h^3$ with mass, $m_P = 60\,h^3 \rho_P$.

For a nebula of mass $0.1\,M_\odot$ with a 2% solid component, in the form of a disk of radius, say, 40 au, the mean surface density, i.e. mass per unit area of disk, of solids, σ, is $35\,\text{kg}\,\text{m}^{-2}$. For the critical density, as given by (14.19), which varies with distance from the star we can calculate h from (14.20) and hence the mass of the planetesimals, assuming a constant areal density. The masses and dimensions of the resulting condensations, with solid material of mean material density, $\rho_{\text{sol}} = 2,000\,\text{kg}\,\text{m}^{-3}$, are given in Table 14.2.

Table 14.2. The masses, M_P, and radii, R_P, of planetesimals, according to Safronov (1972) in the vicinity of the Earth and Jupiter.

	Earth	**Jupiter**
ρ_{cr} (kg m^{-3})	2.83×10^{-4}	2.01×10^{-6}
$h = \sigma/\rho_{cr}$ (m)	1.24×10^{5}	1.74×10^{7}
$M_P = 60h^3\rho_{cr}$ (kg)	3.20×10^{13}	6.35×10^{17}
$R_P = (3M_P/4\pi\rho_{sol})^{1/3}$ (km)	1.6	42

Planetesimals of dimensions from a few kilometres up to, perhaps, 100 km in the outer Solar System are predicted but this conclusion was challenged by Goldreich and Ward (1973). They showed from thermodynamics principles that the condensations would be smaller — hundreds of metres in extent rather than the more-than-kilometre-size bodies predicted by Safronov. For a solid body of radius 500 m the escape speed is about 0.5 m s^{-1} and the velocity dispersion of the initial Goldreich–Ward planetesimals was estimated as ~0.1 m s^{-1} so that collisions with the largest planetesimals can give accretion to form larger bodies. Eventually planetesimals of the size predicted by Safronov would come about, albeit by a two-stage process.

It has been argued that even a small amount of turbulence in the disk would prevent gravitational instability (Weidenschilling, Donn and Meakin, 1989). The free-fall time for a planetesimal to form from an initial clump would be over a year and if turbulence occurs locally on that timescale then the clump would be stirred up and disperse before it could collapse. These workers suggested that dust grains could adhere to form ever larger structures, ending up as the size of planetesimals. There is some justification for this criticism. A planetesimal clump in the Jupiter region before it began to collapse would have an escape speed ~1.4 m s^{-1} and turbulent speeds of this magnitude would disrupt the clump. However, there is disagreement about whether or not the nebula would be turbulent. Forming planetesimals by gravitational instability is favoured by a non-turbulent disk but angular momentum transfer by the Lyndon-Bell and Pringle mechanism is aided by turbulence.

Theorists in this area claim that, whatever the uncertainties in the actual mechanism for forming planetesimals, they will form on a relatively short timescale so that nearly all the lifetime of the dusty nebula ($\leq 10^7$ years) is available for the next stage, which is forming planets from planetesimals.

14.3.3. *From Planetesimals to Planets*

The basic theory for forming planets from planetesimals is due to Safronov (1972), and subsequent work has been developments, or variants, of it. What he showed was that if the average relative speed of planetesimals was less than the escape speed from the largest of them, then that body would grow and accrete all other local bodies which collide with it. Initially gravitational interactions between planetesimals, equivalent to *elastic collisions* that conserve energy, increase the eccentricity of planetesimal orbits and increase their average relative speed. Eventually this increases the probability of *inelastic collisions* in which energy is lost so that relative speed is damped down. Safronov showed that a balance between these opposing effects occurs when the mean relative speed, v, is of the same order as, but less than, v_e, the escape speed from the largest planetesimal. The escape speed is given by

$$v_e^2 = \frac{2Gm_L}{r_L},$$

so that in general one can write

$$v^2 = \frac{Gm_L}{\beta r_L}, \qquad (14.21)$$

where m_L and r_L are the mass and radius of the largest planetesimal and β is a factor in the range 2 to 5 in most situations.

Assuming that all colliding bodies adhere then the rate of growth of the largest body is proportional to the collision cross section, which depends not only on the physical cross section but also includes the gravitational focusing effect due to the mass of the body. The rate of growth of a spherical body of mass m and radius r accreting bodies much less massive than itself is given by the Eddington accretion mechanism (Appendix F) as

$$\frac{dm}{dt} = \pi r \left(r + \frac{2Gm}{v^2} \right) \rho v, \tag{14.22}$$

where ρ is the mean local density of the material being accreted and v its speed relative to the accreting body, which comes from (14.21).

From (14.21) and the relationship

$$\frac{m}{m_L} = \frac{r^3}{r_L^3}, \tag{14.23}$$

where m and r are the mass and radius of a general planetesimal, (14.22) becomes

$$\frac{dm}{dt} = \pi r^2 \left\{ 1 + 2\beta \left(\frac{r}{r_L} \right)^2 \right\} \rho v. \tag{14.24}$$

From (14.23) and (14.24) the ratio of the *relative rate of growth* of a general body to that of the largest body is

$$\frac{\frac{(dm/dt)}{m}}{\frac{(dm_l/dt)}{m_L}} = \frac{r_L}{r} \frac{1 + 2\beta(r/r_L)^2}{1 + 2\beta}. \tag{14.25}$$

This ratio is unity both when $r = r_L$ and $r = r_L/2\beta$. For intermediate values the ratio is less than unity and the relative size of the two bodies diverges. If there are two planetesimals in which one is just marginally larger than the other then r_L/r is just greater than unity and the larger one grows relatively faster, thus increasing the value of r_L/r. Eventually, when $r = r_L/2\beta$, the ratio of masses remains constant at r_L^3/r^3 or $8\beta^3$. For β between 2 and 5 this corresponds to the mass ratio of the largest forming body to the next largest of between 64 and 1000.

With this background the timescale for the formation of a terrestrial planet or the core of a major planet can be estimated. For a particle in a circular orbit of radius r the speed in the orbit is $2\pi r/P$ where P is the period of the orbit. If the random speed perpendicular to the mean plane of the system is less than or equal to v then the orbital inclinations will vary up to $\phi = vP/2\pi r$. The material at distance r will be spread out perpendicular to the mean plane though a distance $h = 2r\phi = vP/\pi$ so that

$$\rho = \frac{\sigma}{h} = \frac{\pi\sigma}{vP}, \tag{14.26}$$

where σ is the areal density.

From (14.24), for the largest body,

$$\frac{dm_L}{dt} = \pi r^2(1 + 2\beta)\rho v. \tag{14.27}$$

If ρ_{sol} is the density of the material forming the body then $m_L = \frac{4}{3}\pi\rho_{\text{sol}}r_L^3$ and we also have, from (14.26), $\rho v = \pi\sigma/P$. Inserting these values into (14.27) gives

$$\frac{dm_L}{dt} = Am_L^{2/3}, \tag{14.28}$$

where

$$A = \frac{\sigma(1 + 2\beta)}{P}\left(\frac{3\pi^2}{4\rho_{\text{sol}}}\right)^{2/3}.$$

Integrating from $m_L = 0$ when $t = 0$ to the formation time, t_{form} when the planet or core has its final mass, M_p gives the formation time as

$$t_{\text{form}} = \frac{3P}{\sigma(1 + 2\beta)}\left(\frac{4\rho_{\text{sol}}}{3\pi^2}\right)^{2/3} M_p^{1/3}. \tag{14.29}$$

Equation (14.29) assumes that σ is constant although it will actually decrease with time. However, if initial σ values are used then the equation may be used to give an order-of magnitude lower bound to the formation time.

Clearly the formation time depends on the mass of the disk and the distribution of that mass within it. A more massive disk will give larger values of σ but this will introduce the problem of disposal of the surplus disk material. As an illustration we consider a disk of mass $0.1\,M_\odot$, radius 40 au with a 2% solid fraction and a variation of areal density of the form $\sigma = C/R$ where R is the distance from the Sun. If the total mass of the disk is M, the proportion of solids is f and the radius of the disk is S then

$$0.1fM_{\odot} = \int_0^S \frac{C}{R} 2\pi R dR = 2\pi CS \text{ giving } C = \frac{0.1fM_{\odot}}{2\pi S}.$$

At the distance of the Earth the areal density is $707\,\mathrm{kg\,m^{-2}}$ and at the distance of Jupiter, 5.2 au, it is $136\,\mathrm{kg\,m^{-2}}$. With $(1 + 2\beta) = 8$ and $\rho_{\mathrm{sol}} = 5.5 \times 10^3\,\mathrm{kg\,m^{-3}}$ this gives a time for forming the Earth of 7.9×10^6 years. For Jupiter with $\rho_{sol} = 4 \times 10^3\,\mathrm{kg\,m^{-3}}$, the formation time for a 10 $M_{\oplus}$ core is 8.3×10^8 years. The formation time for Neptune ($\sigma = 24\,\mathrm{kg\,m^{-2}}$) is, for any reasonable mass of core, greater than 10^{10} years — which greatly exceeds the age of the Solar System.

The lifetimes of nebula disks are a few million years so the basic Safronov theory indicates that by the time the cores of major planets had formed there would be no nebula gas present to be accreted to form a massive atmosphere. Suggestions have been made drastically to reduce the core-formation times. One way is to have local enhancements of density within which the planets form, but the enhancements would have to be very large. If viscous drag on the planetesimals slows them down then, from (14.22), this would increase the capture cross section, which would help a little. Stewart and Wetherill (1988) proposed what they called a *runaway growth* model that incorporated the previously-suggested modifications of the Safronov theory plus energy equipartition, so that the larger masses move more slowly. This would increase the probability of large masses combining when they come together. With those ideas, taken to the limit, they found formation times from 3.9×10^5 years for Jupiter up to about 3×10^7 years for Neptune — the latter time still greater than the expected lifetime of a circumsolar disk.

Once a major-planet core formed then, assuming the presence of gas in the disk at a reasonable fraction of its original density, the formation time for accreting an atmosphere is of order 10^5 years and presents no tight constraint on theories.

14.4. Migration Mechanisms

From the analysis of the previous section it is clear that Uranus and Neptune, and possibly Jupiter and Saturn, could not have formed in their present locations. The solution to this problem is that they should form closer in, where formation times are well within the disk lifetime, and then migrate outwards to reach their present locations. The time available for them to do this is much greater than the lifetime of the disks since well after the gaseous component of a disk has disappeared there will be a considerable mass of planetesimals in orbit around the Sun, or any other star: interactions of a planet with these can give outward migration.

Here we give a description of the migration mechanisms that can occur (Lubow and Ida, 2011), first those dependent on planet interactions with the gaseous disk and then that dependent on planet-planetesimal interactions.

14.4.1. *Type I Migration*

This type of migration occurs when the planet stays in contact with the disk and applies to planets with mass less than about that of Saturn. The orbit of the planet most strongly affects those elements of the disk that are orbiting with a period that is a simple multiple or fraction of the planet's period. These portions of the disk experience forces that tend to make their orbits eccentric and to line up their periapses, i.e. their closest distances to the star. This produces, both inside and outside the planet's orbit, regions of higher density that take on the form of spiral waves. These regions of higher density affect the planet gravitationally. The outer spiral wave reduces the planet's speed and hence its angular momentum, thus moving it inward. Conversely, the inner enhanced density wave increases the speed and angular momentum, so moving the planet outward. In almost all cases the effect of the outer spiral density wave is the stronger so the planet moves inwards.

14.4.2. *Type II Migration*

A planet massive enough to open a gap in a gaseous disk, one approximately more massive than Saturn, undergoes what is known as Type II migration. When the mass of a perturbing planet is large enough, the tidal torque it exerts on the gas transfers angular momentum to the gas exterior to the planet's orbit, and takes angular momentum from gas interior to the planet's orbit, thereby repelling gas from a region around the planet's orbital path. This creates a low density gap in the disk.

In Section 14.2.3, the Lynden-Bell and Pringle mechanism was described in which inner disk material moved further inwards and outer material further outward due to the occurrence of energy loss in the disk due to viscosity. Any gaps in the disk also follow this pattern of movement and the planet situated in a gap migrates with the gap. Thus a planet in the inner part of the disk migrates inward and one in the outer part of the disk migrates outward.

14.4.3. *The Interaction of a Planet with Planetesimals*

For a disk of mass 0.1 $M_\odot$ with, say, 2% of solid material, one half of which is contained in planetesimals, the total mass of planetesimals is about the mass of Jupiter, or just over 300 $M_\oplus$.[1] Not much of this, perhaps 30 $M_\oplus$ will be needed to produce terrestrial planets and the cores of major planets and the remainder will be left behind when the gaseous part of the disk dissipates with an average planetesimal mass of 10^{15} kg there will be of order 10^{12} planetesimals, on or near the mean plane of the original disk, spread over the area of the disk following a density distribution closely proportional to that of the original disk. This material is clearly going to affect the orbits of the planets moving within it.

In a system with two bodies in commensurate orbits, i.e. with a ratio of orbital periods that is a ratio of small integers, the inner body adds energy to the outer one and the outer diminishes the energy of the inner one. Thus a planet in a sea of planetesimals will be gaining energy from those in interior commensurate orbits and losing energy to those in exterior commensurate orbits. The net effect,

[1]The symbol $\oplus$ indicates 'Earth'.

whether migration is inward or outward, depends on the distribution of planetesimals.

14.4.4. *The Nice Model*

In 2005, four astronomers — Rodney Gomes, Hal Levison, Alessandro Morbidelli and Kleomenis Tsiganis, based at the Observatoire de la Côte d'Azur in Nice, produced a simulation for the evolution of the four solar-system major planets in the presence of a disk of planetesimals of total mass between 30–50 $M_{\oplus}$ It is assumed that all planetesimals within the region initially occupied by the planets would have either been absorbed or ejected from the Solar System so the planetesimal disk had an inner boundary just outside the initial planetary region and stretched out to between 30 to 35 au from the Sun. The areal density of planetesimals was taken to fall inversely with heliocentric distance and was sharply terminated at the outer limit.

The gas giants were initially placed in near-circular orbits with semi-major axes 5.54 au for Jupiter and 8.65 au for Saturn, giving close to, but not at, a 2:1 period commensurability Several simulations were run with different positions for the ice giants. Neptune's orbital radius was between 11 and 13 au and that of Uranus between 13.5 au and 17 au, with the proviso that the radii should differ by at least 2 au. Having Neptune closer in than Uranus, differs from the present arrangement. There were between 1,000 and 5,000 simulated planetesimals, which gave a satisfactory representation of the statistical behaviour of the model. In the simulations the planet interacted gravitationally with the planetesimals but the planetesimals did not interact with each other. Forty-three simulations were run and the results were subjected to statistical analysis. There was a wide variety of outcomes, but some general patterns of behaviour became evident.

A typical output is shown in Figure 14.6. The three lines shown for each planet are the aphelion, Q, the perihelion, q, and the semi-major axis, $(Q + q)/2$. The eccentricity of the orbit is given by $e = (Q - q)/(Q + q)$ — shown on the right-hand side adjacent to the double-arrowed line.

For about 6 million years not much seems to be happening but Saturn then reaches a 2:1 ratio of its orbital period to that of Jupiter,

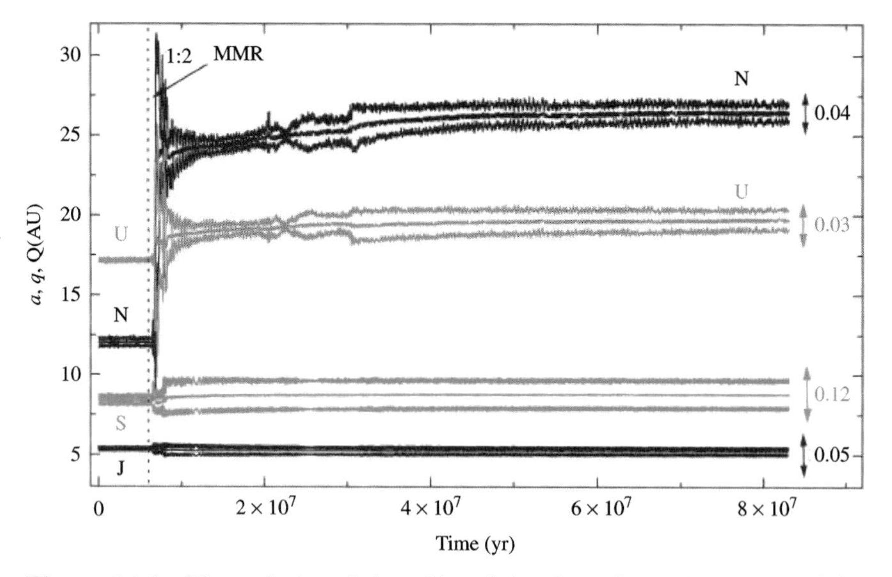

Figure 14.6. The evolution of the orbits of the giant planets in one run of the Nice model.

shown as mean motion resonance (MMR) in the figure. The resonance condition increases the orbital eccentricity of Saturn that takes it close to Neptune, which is propelled beyond the orbit of Uranus, which is also perturbed into a more eccentric orbit. Now both Uranus and Neptune are in the planetesimal region and due to what is known as *dynamical friction* the cloud of planetesimals has the same action as a gas in rounding off the orbits of the ice giants. When the orbits of the ice giants have rounded-off within the planetesimal region they are further influenced by interactions with the planetesimals, the net result of which is to give a gradual drift outwards.

Planetesimals ejected inwards into the gas-giant region cause an increase in the rate of orbital evolution of those planets. After 80 million years the system has settled down; the indicated eccentricities are of the right order of magnitude, the greatest discrepancy being that of Saturn that is actually 0.056.

Of the 43 trial runs of the Nice model 14 were 'unsuccessful in that they gave outcomes very different from the present Solar System. The 29 'successful' runs were divided into two classes. Class

A consisted of runs, in which only the ice giants interacted and class B in which Saturn interacted with one or both of the ice giants. For both the classes, planets ended up in the correct order in terms of distance from the Sun and with small eccentricities. Class B outcomes were somewhat better in terms of the present orbital characteristics of the Solar System.

Figure 14.7 gives a summary of the outcomes for the successful runs. For each of the classes the averages of the eccentricities, orbital inclinations and semi-major axes were found together with the standard deviation of the values for each quantity. Class A results are shown as open grey circles, class B as open black circles and true values by solid black circles. The inclinations are measured relative to the orbit of Jupiter — hence Jupiter's inclination is taken as zero. Solar-system planet inclinations are normally given relative to the ecliptic, the plane of the Earth's orbit, in terms of which the inclination of Jupiter is just over 1°.

The results for class A (15 runs) and class B (14 runs) are both quite good with class B giving closer agreement, especially for the semi-major axis of Uranus. The vertical and horizontal bars give the standard deviation for each quantity for each class and it will be seen that most averages are within one standard deviation of the true value.

The possible disruption of the satellite families of the ice giants was investigated. Both ice giants were given a set of satellites similar to that of Uranus. Only four of the successful runs disrupted the satellite families to any significant extent. *Regular satellites*, those closest to the planets in nearly circular orbits close to the planet's orbital plane, survived for nearly all successful simulations but outer *irregular satellites*, thought to be captured bodies, were mostly lost. The irregular outer satellites would have to have been captured after the close interactions.

The Nice model has a high level of acceptance by the astronomical community. If it has a problem, it is that of justifying the starting point since the initial orbital radii are well outside the distances suggested by the basic Safronov theory and the expected lifetimes of disks. However, there are mechanisms for outward migration while

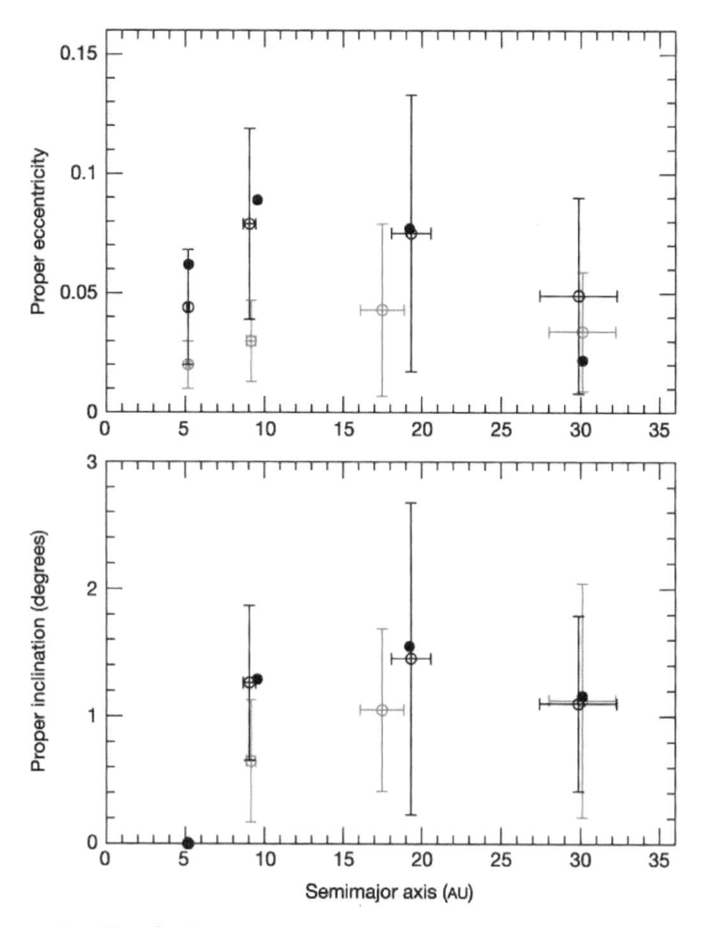

Figure 14.7. The final outcomes for average semi-major axes, eccentricities and inclinations, and their standard deviations, of class A (grey open circles) and class B (black open circles) simulations. The true values of semi-major axis, eccentricity and inclination are shown by the solid black circles.

the disk is present, and the action of inner planetesimals, before the initial stage of the Nice model, could also have moved planets outwards from their initial positions.

14.5. The Proportion of Stars with Planets

Estimates of the proportion of stars with planets vary with time and tend to increase; a recent estimate, based on reasonable premises, is 0.34 (Section 13.5). On the basis that a process similar to that proposed by Laplace takes place, where a disk is left behind by a collapsing nebula, modified by the Lynden-Bell and Pringle mechanism that transports angular momentum outwards, it is likely that the majority of young stars would develop a circumstellar disk. How this evolves would depend on its extent and mass, the statistics of which is unknown, so it is difficult to assess what proportion of stars would produce planets. However, it seems likely that the observational estimate could be explained by this model of planet formation.

14.6. Smaller Bodies of the Solar System

Many solid solar-system bodies are covered with collision scars, such as mare basins and craters on the Moon. It has been deduced that about 500 million years after solar-system formation the bodies in the inner Solar System were subjected to a period of what is known as the late heavy bombardment (LHB). It has been related to the time when Uranus and Neptune penetrated into the planetesimal region and when many planetesimals were projected inwards. However, the timing of this event does not seem compatible with the output of the Nice model, as seen in Figure 14.6, which suggests that the Solar System would have settled down much earlier but a modification of the Nice model might resolve the timing discrepancy.

14.6.1. *Asteroids*

The planetesimals thrown inwards by Uranus and Neptune would not only have collided with bodies in the inner Solar System but would also have interacted with them gravitationally and collided with each other. Some of these bodies would have ended up in stable heliocentric orbits, particularly in the region between Mars and Jupiter, well away from those bodies, and constitute the present Asteroid Belt. There are tens of thousands of bodies in the Asteroid Belt but their total mass is just 4% of the mass of the Moon so they are just that tiny fraction of the total mass of planetesimals that happened to end up in 'safe' orbits. The Nice-model simulations also produced some planetesimals as Trojan asteroids that are observed accompanying Jupiter and Neptune. These are asteroids that are in virtually in the same orbits as the planets but are either 60° ahead or 60° behind the planet in its orbit. These are stable positions for a small body under the combined gravitational effects of the Sun and planet — stable in that if they depart from the stable position the combined forces of the Sun and planet push them back to where they came from.

14.6.2. *The Kuiper Belt*

The original distribution of planetesimals extended beyond the present orbit of Neptune, the orbital radius of which is 30 au. Those scattered inward by Neptune that passed close to the Sun would appear as comets due to vaporization of their volatile content when heated. This still happens producing *short-period comets*, although at a comparatively low rate. These are comets that orbit mostly within the planetary region. At each perihelion passage they lose some of their volatile content and after about 1,000 orbits they become dark inert objects, something like low-density asteroids. A well-known short-period comet is Halley's Comet, with a period of about 75 years.

In 1951, the Dutch, later American, astronomer, Gerard Kuiper, predicted that there would be a population of small bodies beyond the orbit of Neptune. This population of small bodies is now known to exist and it is estimated to have about 20 times the width and 200 times the mass of the Asteroid Belt.

14.6.3. *Dwarf Planets*

Dwarf planets are defined as bodies in heliocentric orbits massive enough to be both spherical, or nearly so, under the influence of self-gravitational forces and to clear all other objects from their paths. There are six such bodies known at the time of writing, five within the Kuiper Belt — Pluto, Eris, Haumea, Makemake and V774104, and the other is Ceres, a body of diameter roughly 1,000 km, that is within the Asteroid Belt. Dwarf planets are interpreted as aggregations of planetesimals that did not grow to full planetary size.

Some Kuiper-belt objects have orbital periods commensurate with that of Neptune, an outstanding example being Pluto. The present ratio of periods is 1.504 but Pluto's orbit changes with time and the time-average of the ratio is 1.5. At perihelion Pluto's orbit just ducks inside that of Neptune. Although the *orbits* closely approach each other the planets never do so; the commensurability of the orbits plus the 17° inclination of Pluto's orbit, keeps them apart. Actually Pluto approaches Uranus more closely than it approaches Neptune.

14.6.4. *The Oort Cloud*

At a distance of tens of thousands of au, and so reaching a considerable fraction of the way to the nearest stars, there is a cloud of comets known as the *Oort Cloud.* They are taken as planetesimals that were ejected to their present locations by interactions with Jupiter. Their orbits are perturbed by occasional passages of stars close to, or through, the cloud and if they then pass close to the Sun they are seen as *new comets.* The term 'new' comes from the fact that, coming from so far away, they have eccentricities close to unity and large semi-major axes, so the intrinsic energy of their orbits, i.e. energy per unit mass, is only just negative, meaning that they are weakly bound to the Solar System. When they pass through the planetary region, perturbation by planets changes in their intrinsic energy by an amount that can be either positive or negative, and is usually larger in magnitude than their original intrinsic energies. If the change is negative then their final intrinsic energy is much less than the original energy and the new orbit will be smaller and much more tightly bound. If the change is positive then the final intrinsic energy will almost certainly be positive and hence they will leave the Solar System. A consequence of this scenario is that it is almost certain that they have never been seen before as objects coming from the Oort Cloud and for this reason they are called *new comets.*

14.7. The Inclinations of Exoplanet Orbits

In Section 13.4.2, the observations of SOMs were described showing that many exoplanets were in retrograde orbits. With a star and planets both derived from the same nebula it would be expected that the SOMs would be zero, or at least small allowing for perturbation of some kind. The initial reaction to the observations, as illustrated in Figure 13.8, is that this presented a considerable challenge to the NT. However, several mechanisms have been suggested to resolve the apparent difficulty.

If there is an external star then it can perturb a planet's orbit by the Kozai effect (Kozai, 1962). This involves an interplay between the eccentricity and the inclination of the orbit and can give an SOM up to 40°, starting with zero SOM. (Fabrycky and Tremaine, 2007). Another effect of an external star is that the circumstellar disk, in which planets form, could either be changed in orientation or be warped by stellar perturbation.

Another mechanism involves interactions of pairs of exoplanets. Because of different rates of orbital decay, planets may interact closely and have the inclinations of their orbits greatly affected, even to the extent of complete reversal of motion into a retrograde orbit. Finally, Rogers, Lin and Lau (2012) suggested that the effect of internal gravity waves — wave-like undulations of interior material due to gravitational disturbance — can cause external layers of the star to spin round a different axis from that of the main bulk of the star. On this interpretation large SOM angles do not really occur. The spin of the great bulk of the star is perpendicular to the orbital plane of the planets.

14.8. Exoplanets Around Binary Stars

The observation of an exoplanet around the more massive star in the binary system γ-Cephei (Section 13.5) was suggested as a problem for the NT because in the environment of the orbiting stars the local medium would be stirred up, which would prevent a planet forming, However, as explained in Section 15.3, there are periods in a disk-forming region where the star-number density can be very large and where interactions between stars can occur. It is possible that an interaction can lead to a planet around one star to be captured by another star, which could be a member of a binary system.

Another potential scenario is a two-stage process in which a planet is removed from one star by a close interaction with another star and the escaped planet is then captured by a third star that, again, might be a member of a binary system.

These kinds of capture scenario could also explain planets at large distances from stars, which is difficult to explain by the conventional NT planet-forming process.

14.9. Satellite Formation

In the process of capturing gas to form an atmosphere it is envisaged that though various processes, including the Lyndon-Bell and Pringle mechanism, a circumplanetary disk would form. Then a small-scale version of the process for forming planets could take place to produce satellites.

The theory closely follows that for the formation of planets, given in Section 14.3. but with very different parameters — for example, the circumplanetary disk has a much greater areal density than a circumstellar one. A treatment of satellite formation in a circumplanetary disk is given in Section 15.12 that, in a slightly modified form with different parameters, can be applied to the situation engendered by the NT.

This theory explains the existence of *regular satellites* that have the characteristics that they are in direct close-to-circular orbits close to the equatorial plane of the planet and are ones nearest to the planet. There are other *irregular satellites* that do not have all these characteristics and are explained as captured bodies.

14.10. An Overview of the Nebula Theory

Since its introduction in the early 1960s, the NT has constantly been challenged by those making new observations, Thus, although the observation of disks around new stars was a boost for the NT, the comparatively short lifetime seemed inconsistent with the timescales that theory indicated to form the cores of major planets in their present locations, Other observations that were indicated as a problem for the NT were the detection of exoplanetary orbits of high inclination or even retrograde. The existence of a planetary companion for one member of a binary system also seemed to be a problem because the disturbed state of the medium in the presence of orbiting binaries might inhibit the formation of planetesimals.

In all these cases quite plausible solutions were found for the apparent problems. The NT is the 'standard model' of planetary formation, i.e. the one accepted by the great majority of astronomers and work continues on developing various aspects of the theory.

Problems 14

14.1 A star with mass 2.5×10^{30} kg, moment of inertia factor 0.06, and radius 8.0×10^5 km has a magnetic dipole moment $10^{25}\,\mathrm{T\,m^3}$ and is losing mass at a rate $10^{13}\,\mathrm{kg\,s^{-1}}$ at a speed of $6 \times 10^5\,\mathrm{m\,s^{-1}}$.

(i) Out to what distance, from the centre of the star, in units of the stellar radius, will the lost material co-rotate with the star?

(ii) If the original angular spin rate was 2.4×10^{-4} radians $\mathrm{s^{-1}}$ then what is the spin rate when the mass of the star is 2.49×10^{30} kg, assuming the rate of loss of mass is constant? ($\mu_0 = 1.2566 \times 10^{-6}\,\mathrm{m\,kg\,s^{-2}\,A^{-2}}$)

14.2 Using the basic Safronov equation (14.29), with $\beta = 4$, find the time for the formation of

(i) Venus with $P = 0.615$ years, $\sigma = 700\,\mathrm{kg\,m^{-2}}$, uncompressed density $\rho_{sol} = 4.4 \times 10^3\,\mathrm{kg\,m^{-2}}$ and $M_P = 4.868 \times 10^{24}$ kg.

(ii) A Saturn core with $P = 29.46$ years, $\sigma = 50\,\mathrm{kg\,m^{-2}}$, $\rho_{sol} = 4.1 \times 10^3\,\mathrm{kg\,m^{-3}}$ and $M_{core} = 4. \times 10^{25}$ kg.

Chapter 15

The Capture Theory

15.1. Introduction

At the same time that the first ideas were developing for a revised Nebula Theory another model for planet formation was being developed of a completely different kind. This is the Capture Theory (CT) that has changed significantly since it was first presented by Woolfson (1964). It has been modified and adapted in the light of new observations about exoplanets, star-forming clouds and the Solar System, as they occurred. While it has been influenced by new information, during its development it has sometimes made assumptions that have later been confirmed by observations. It is a *dualistic theory*, meaning that the material for stars and planets come from separate sources. It also describes a chain of events for planet formation, each causally related to what preceded it, that naturally leads to a proposed event, the consequences of which explain all the important features of the Solar System.

The development of the CT has embraced the whole range from the formation of stars (Section 4.8) to the explanation of details of the contents of the Solar System.

15.2. Observations Relating to Star Formation

In Figure 4.2, some theoretical development paths for YSOs were shown on an H–R diagram for different stellar masses. They occupy the same regions of the H–R diagram as old stars leaving the main sequence. However, if for a stellar cluster there are stars occupying the very large mass region of the main sequence then it is certain that the cluster, and hence the stars in it, are young. Williams and Cremin (1969) studied YSOs in four young galactic clusters. From its position on the H–R diagram the age and mass of an YSO can be deduced. Figure 15.1 shows the results for the cluster NGC2264, from which it was concluded that:

(i) The first stars were produced about 8×10^6 years ago;
(ii) The first stars have an average mass somewhat greater than $M_\odot$;
(iii) There are two streams of development, one with reducing mass with time and the other, starting about 5 million years ago, with increasing mass with time;
(iv) The overall rate of star formation increases with time.

The stream with reducing mass can be identified as due to the increasing density of the gas in a collapsing cloud, reducing the Jeans critical mass and allowing stars of reducing mass to form. The stream of increasing mass can be related to the model of Bonnell, Bate and Zinnecker (2005) for the formation of massive stars by the

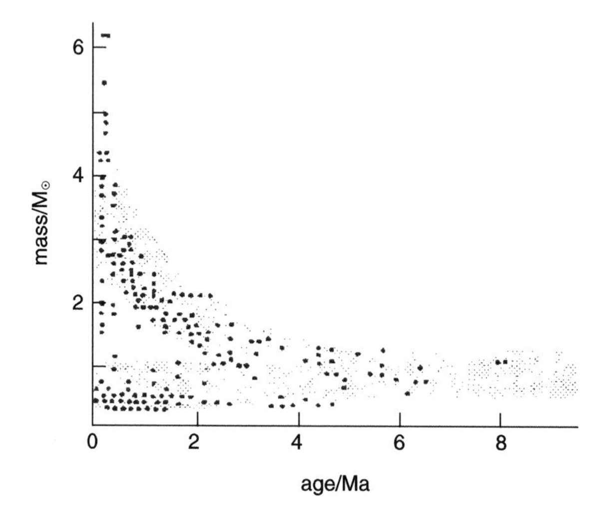

Figure 15.1. The masses of stars produced in a young stellar cluster as a function of time. The origin represents ‘now’ (After Williams and Cremin, 1969).

aggregation of smaller-mass stars or protostars in a dense stellar environment. Very large mass stars cannot form directly because at a certain stage, when a very hot core forms as the protostar collapsed, the radiation from it would prevent the further accretion of material.

15.3. Interactions in a Star-Forming Cloud

In Sections 4.7 and 4.8, there was described the formation of DCCs and the subsequent formation of protostars, later collapsing to form, first YSOs and then main-sequence stars. This process, while gas is still present and is producing a stellar cluster with increasing numbers of stars, is said to be in an *embedded state*, meaning that the stars are still embedded in gas. Eventually the collapse of the cluster will first slow down and finally cease as the gas pressure builds up and for a period of 5 million years or so the cluster will be in a *dense embedded state* (DES) where the stellar number density (SND) can be very high. Gas is slowly expelled by stellar radiation but, after a few million years, massive-star supernovae increase the rate of gas expulsion. As the gravitational influence of the gas is removed the stellar cluster begins to expand. In 90% of cases the expansion continues indefinitely to give field stars and field binary systems. In the other 10% of cases a galactic stellar cluster forms, typically containing a few hundred stars.

The maximum SND maintained during the DES can be very high; for the Trapezium Cluster, within the Orion Nebula, the core SND is estimated to be several times $10^4\,\mathrm{pc}^{-3}$. Bonnell, Bate and Vine (2003), in a simulation using high-definition SPH, showed that during the DES, of duration 5 million years, the cloud fragmented, each fragment, containing tens of stars, with SND about $2\times10^5\,\mathrm{pc}^{-3}$, although the whole-cloud average SND was about two orders of magnitude less. Subsequently the fragments expanded and combined to form larger fragments that moved closer together; the peak fragment SND decreased but the whole-cloud average increased. Finally there were 400 stars in five fragments with the maximum fragment SND somewhat greater than $2\times10^4\,\mathrm{pc}^{-3}$ and whole-cloud SND peaking at $2\times10^4\,\mathrm{pc}^{-3}$.

We now consider what would happen within this period of high SND. A typical newly-formed protostar, of mass about $0.5M_\odot$, could have radius 2000 au, density $10^{-14}\mathrm{kg\,m^{-3}}$ and temperature 20 K. The free-fall time to high density for such a body is $t_{ff} =$ 21,000 years; after a time $0.8t_{ff}(\sim$ 17,000 years) its radius will have fallen to 1,000 au. The average stellar speed in a DES is about $1\,\mathrm{km\ s^{-1}}$ (Gaidos, 1995) so in 17,000 years, while an extended object, a protostar can travel more than 3,000 au. For an SND of $2 \times 10^4\,\mathrm{pc^{-3}}$ the average interstellar distance is just over 8,000 au. From protostar dimensions, the distances they travel and the SNDs that occur it is clear that close approaches of extended protostars and condensed stars will take place. We now consider a tidal interaction between an extended protostar and a condensed star. The frequency of such interactions is considered quantitatively in Section 15.5.

15.4. Capture-Theory Simulations

The standard model of planet formation for two decades was that presented by Jeans (1917) who proposed a theory that the Solar System formed from the passage of a massive star past the Sun. A filament of matter, drawn out of the Sun, was gravitationally unstable and broke up into a string of blobs that eventually collapsed to form planets. The blobs, attracted by the retreating massive star, were left in orbit around the Sun.

This theory introduced new ideas such as tidal effects and the gravitational instability of gaseous filaments, for both of which Jeans produced very sound theoretical treatments. Similar mechanisms occur in the CT but in a completely different context. A CT simulation by Oxley and Woolfson (2003, 2004), using SPH plus radiation transfer, is shown in Figure 15.2. The parameters for the simulation were as follows:

Star characteristics

Mass, $M_* = 2 \times 10^{30}\,\mathrm{kg} \approx M_\odot$
Luminosity, $L_* = 4 \times 10^{26}\,\mathrm{W} \approx L_\odot$.

Protostar characteristics

Mass, $M_P = 7 \times 10^{29}$ kg $\approx 0.35\, M_\odot$
Initial radius, $R_P = 800$ au
Initial temperature, $T_P = 20$ K
Mean molecular mass of material, $\mu = 4 \times 10^{-27}$ kg.

Characteristics of the protostar orbit

Initial distance of centre of protostar from star, $D = 1{,}600$ au
Closest approach of protostar orbit to star, $q = 600$ au
Eccentricity of the orbit, $e = 0.95$.

Almost the whole protostar is tidally distorted into a dense filament that, as Jeans showed theoretically, is gravitationally unstable and breaks up into a string of blobs that have greater than the Jeans critical mass and collapse (Figure 15.3). Five of the blobs, with masses $4.7\, M_J, 7.0\, M_J, 4.8\, M_J, 6.6\, M_J$ and $20.5\, M_J (M_J$ is mass of Jupiter}, are captured into orbits around the star — hence the name of the theory. Some blobs escape into interstellar space as *free-floating planets*, bodies of planetary mass that have been observed within the ISM (Lucas and Roche, 2000; Sumi *et al.*, 2011).

There is one other way in which a CT process may occur. In Section 4.8 the formation of protostars was described as due to the collision of turbulent gas streams within a collapsing DCC. If such a collision takes place close to a condensed star then this can, under suitable conditions, also give planet formation. In Figure 15.4 each stream has mass $0.5\, M_\odot$, density 4×10^{-15} kg m^{-3} and speed 1 km s^{-1}. The condensations A, B and C, with masses $1.0\, M_J, 1.6\, M_J$ and $0.75\, M_J$ respectively, are captured.

The CT mechanism is very robust and gives planet formation at scales from one-tenth to ten times the scale used in Figures 15.2 and 15.4.

The planet masses for the two CT simulations are within the range found for exoplanets but the initial orbits are not. For the eight captured protoplanets in Figures 15.2 and 15.4 the semi-major

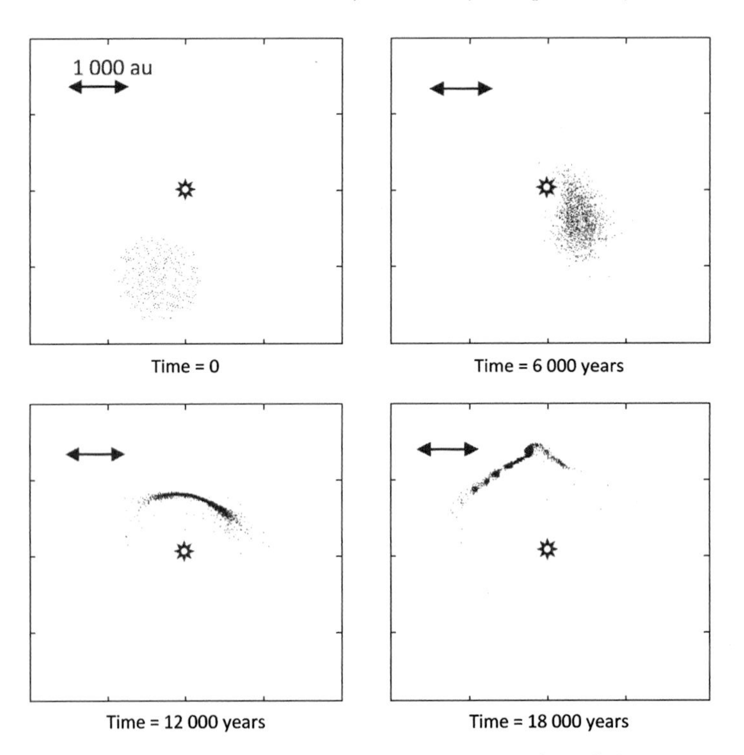

Figure 15.2. An SPH simulation with radiation transfer, of the interaction of a star and a protostar (Oxley and Woolfson, 2004).

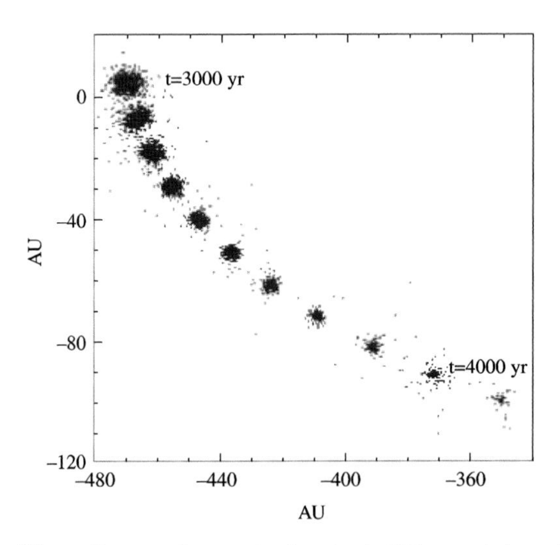

Figure 15.3. The collapse of a protoplanet at 100-year intervals (Oxley and Woolfson, 2004).

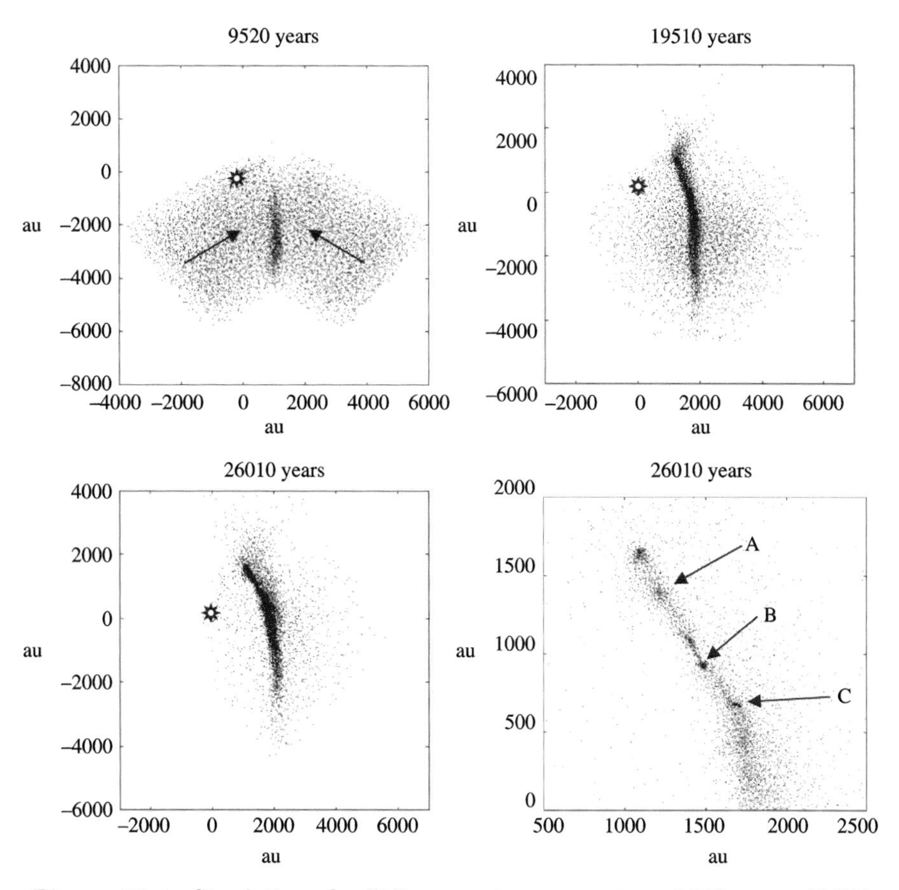

Figure 15.4. Simulation of colliding gas streams at times 9,520 years, 19,510 years and 26,010 years. The final frame shows a higher resolution view at 26,010 years (Oxley and Woolfson, 2004).

axes, a, and eccentricities, e, are as follows:

$$(a, e) = (1{,}247\,\text{au}, 0.835) \qquad (a, e) = (1{,}885\,\text{au}, 0.7725)$$

$$(a, e) = (1{,}509\,\text{au}, 0.765)$$

$$(a, e) = (1{,}325\,\text{au}, 0.736) \qquad (a, e) = (2{,}686\,\text{au}, 0.902)$$

$$(a, e) = (4{,}867\,\text{au}, 0.768)$$

$$(a, e) = (1{,}703\,\text{au}, 0.381) \qquad (a, e) = (1{,}736\,\text{au}, 0.818)$$

These orbits are more extensive than any found for exoplanets and their eccentricities are large, although some exoplanet orbits do have large eccentricities.

The CT process of planet formation is described as 'top-down', in which major planets are produced directly from an above-Jeans-critical-mass of dusty gas — the condensations seen in Figures 15.2, 15.3 and 15.4. Smaller bodies, such as occur in the Solar System, require that one or more of the initial planets must break up. From Figures 15.2 and 15.4 the timescale for CT planet formation is of order 10^4 years. This contrasts with the NT process, which is termed 'bottom-up' and involves four stages in which objects of increasing size are produced. The timescale for the NT multi-stage process of producing planets is of order 10^6 years in the terrestrial region of the Solar System, but increases rapidly with increasing distance from the Sun.

15.5. The Proportion of Stars with Planets

A soundly-based estimate of the proportion of stars with planets is 0.34 (Borucki *et al.*, 2011), and a plausible theory should be able to explain a proportion of this order of magnitude.

Woolfson (2016) developed a model to quantify the frequency of interactions between protostars and condensed stars in a DES and hence estimated the proportion of stars with planets. The model was investigated for SNDs of $n = 5{,}000\,\mathrm{pc}^{-3}$ to $25{,}000\,\mathrm{pc}^{-3}$ in steps of $5{,}000\,\mathrm{pc}^{-3}$ and initial protostar radii $R_P = 1{,}000, 1{,}500$ and $2{,}000$ au with corresponding free-fall times of $t_{ff} = 10{,}200, 18{,}800$ and $28{,}900$ years for a protostar of mass $0.3\,M_\odot$. The motion of a collapsing protostar among the stars is followed for a maximum period $0.8t_{ff}$, at which stage it has about one half of its original radius.

A random configuration of N_S stars in a cubical cell is surrounded by 26 'ghost cells', each containing a similar configuration. When a star leaves the central cell it re-enters at the same position on the opposite face, so keeping the density constant within the central cell. The length of the cell side, a, is set to give the required SND and stars in the ghost cells are only included out to a distance ma from the centre of the central cell, so the spherical cluster contains, on average, N_T stars where

$$N_{\mathrm{T}} = 4\pi m^3 N_{\mathrm{S}}/3. \tag{15.1}$$

The protostar's motion is affected by the stars in the spherical cluster plus that of the residual gas in the fragment, the mass of which is kept constant. The gas has two effects — firstly, gravitational and, secondly, by exerting a drag effect on bodies so reducing their speeds.

Many CT simulations of the type shown in Figure 15.2 indicate that if the closest approach of a protostar orbit, r_c, to a solar-mass star was between 0.5 and 1.5 of the protostar radius then a CT event was almost inevitable. Since the CT is based on tidal effects, for a star of mass M_* this gives the condition

$$0.5\left(\frac{M_\odot}{M_*}\right)^{1/3} R_P \leq r_c \leq 1.5\left(\frac{M_\odot}{M_*}\right)^{1/3} R_P, \tag{15.2}$$

where R_P is the current radius of the collapsing protostar.

The masses of the stars are chosen randomly from a distribution with mass index ~ 2.3. (Kroupa, 2001), i.e.

$$f(M) \propto M^{-2.3}, \tag{15.3}$$

with masses in the range 0.5–3.0 $M_\odot$, all greater than the protostar mass. The average mass of the selected stars is 1.00 $M_\odot$.

For gravitationally-interacting stars in a bound region in equilibrium, and with no other forces acting, the Virial theorem (Section 2.2) is valid. This links the translational kinetic energy, K, of the stars to their total potential energy, Ω, by

$$K = -0.5\Omega. \tag{15.4}$$

However, due to gas-dynamical friction, the stars in a DES will have sub-virial energy (Indulekha, 2013) so that

$$K = -\beta\Omega, \tag{15.5}$$

with $\beta < 0.5$. Protszkov *et al.* (2009) have taken β in the range 0.04 to 0.15, corresponding to root-mean-square speeds between 28%

and 55% of the virial value. Stars, and the protostar, were given speeds corresponding to the equipartition value in randomly-chosen directions. Thus with N_T stars in the cluster the ith star has a speed

$$V_i = \left(-\frac{2\beta\Omega}{N_{\mathrm{T}}M_i}\right)^{1/2}. \tag{15.6}$$

The speeds assigned to stars of the extreme masses, $0.5\,M_\odot$ and $3.0\,M_\odot$ as a function of the SND are, with $\beta = 0.04$,

$$\text{For } n = 5{,}000\,\text{pc}^{-3} \text{ and } M_i = 0.5\,M_\odot V_i = 483\,\text{m s}^{-1}$$

$$\text{For } n = 5{,}000\,\text{pc}^{-3} \text{ and } M_i = 3.0\,M_\odot V_i = 198\,\text{m s}^{-1}$$

$$\text{For } n = 25{,}000\,\text{pc}^{-3} \text{ and } M_i = 0.5\,M_\odot V_i = 632\,\text{m s}^{-1}$$

$$\text{For } n = 25{,}000\,\text{pc}^{-3} \text{ and } M_i = 3.0\,M_\odot V_i = 258\,\text{m s}^{-1}$$

For comparison, Gaidos (1995) gave stellar speeds in a DES from 500 m s^{-1} to 2,000 m s^{-1}, corresponding to a larger value of β and allowing larger values of n.

The equations of motion are numerically solved for the N_S stars plus protostar in the basic cell. The gravitational effects of ghost-cell stars are taken into account but they are not moved during each integration step. The gravitational acceleration due to the gas, of mass M_T, within the spherical cluster on a star at vector position $\mathbf{r}$ is given by

$$\mathbf{a}_r = \frac{GM_{\mathrm{T}}}{(ma)^3}\mathbf{r}. \tag{15.7}$$

The integration was carried out using the 4-step Runge–Kutta method. If at the end of any timestep the motion of the protostar gives a periastron distance from a star between 0.5 and 1.5 of its current radius, which changes with time, then planet formation is deemed to have occurred but, if it moves closer, then it is taken that it is disrupted without planet formation. If the planet formation or protostar disruption stage is reached then the trial is terminated and the next one begun, otherwise the calculation is terminated at time $0.8t_{ff}$. At the end of each integration timestep any stars, including the protostar, that have left the basic cell are reintroduced into the cell as previously described and the ghost cells and the cluster fragment are redefined. For four sets of conditions, with $\beta = 0.04$, 1,000 Monte Carlo trials were run with

$$\text{A } N_S = 8,\ m = 1.5,\ \text{giving } N_T = 113$$

$$\text{B } N_S = 23,\ m = 1.0,\ \text{giving } N_T = 96$$

$$\text{C } N_S = 10,\ m = 1.0,\ \text{giving } N_T = 42$$

All these had the ratio of mass of gas to stars equal to 1. Set D is as for C but with mass ratio 3. The percentages of capture events are given in Table 15.1; these results will underestimate the number of CT events because:

(a) Capture-theory events of the type shown in Figure 15.4 have not been included.
(b) The SNDs have been considered up to $2.5 \times 10^4\,\text{pc}^{-3}$. Maximum values up to $10^6\,\text{pc}^{-3}$ have been suggested by McCaughrean and Stauffer (1994) and up to $2 \times 10^5\,\text{pc}^{-3}$ by Bonnell, Bate and Vine (2003).
(c) Some authors have suggested protostar radii greater than 2,000 au.
(d) The value taken for β — 0.04 — gives root-mean-square speeds less than those indicated by Gaidos (1995). Increasing β gives more stars with planets.

The initial protoplanet orbits from CT simulations are very extended and the newly-formed planets will spend a large proportion of their period near apastron so that initial planetary systems can be disrupted by stellar perturbation before their orbits have completely evolved. Woolfson (2004b). has carried out simulations that suggest that up to 20% of planetary systems can be lost in this way. Nevertheless, from the results given in Table 15.1 it is concluded that the CT can explain the estimated proportion of stars with planets.

The present analysis does not directly give the proportion of stars with planets but rather the proportion of *protostars giving planets.* The following example gives an idea of how to transform from the latter quantity to the former. If 100 protostars give 25 sets of planets

Table 15.1. Percentage of CT interactions giving planets with various R_P and n.

R_p(au)	n	Set A Capture (%)	Set B Capture (%)	Set C Capture (%)	Average Capture (%)	Set D Capture (%)
1,000	5,000	0.7	0.6	0.3	0.5	0.3
	10,000	1.8	1.6	1.0	1.5	0.8
	15,000	3.4	2.1	1.8	2.4	1.9
	20,000	3.8	3.3	2.9	3.3	3.2
	25,000	5.3	4.5	4.6	4.8	5.3
1,500	5,000	2.3	2.3	2.7	2.4	2.4
	10,000	6.5	5.6	5.9	6.0	5.5
	15,000	11.6	8.9	10.2	10.2	11.0
	20,000	11.5	13.5	11.9	12.3	12.9
	25,000	15.9	16.3	17.1	16.4	19.5
2,000	5,000	7.9	6.6	5.0	6.5	5.2
	10,000	17.7	16.9	16.7	17.1	17.0
	15,000	29.0	25.9	24.7	26.5	29.1
	20,000	36.4	37.8	33.6	35.9	40.8
	5,000	48.0	46.0	44.2	46.1	52.4

and 10 are disrupted then we have added 25 planetary systems (to pre-existing stars) and 90 new stars to the cluster. The ratio of the number of planetary systems produced to the number of stars added is then $0.25/0.90 = 0.28$.

15.6. Angular Momentum in the Solar System

The failure of the Laplace theory to explain the partitioning of mass and angular momentum in the Solar System was what led to its demise. However, the interaction between the solar wind and the Sun's magnetic field can reduce the angular momentum of the Sun regardless of the way that body formed. As long as the Sun is rotationally stable when it first forms the slow spin of the Sun presents no difficulties for any theory.

The real angular momentum problem is to explain the present locations of the planets For the NT it is to displace the planets outwards from positions much closer to the Sun, i.e. to produce outwards migration and a gain of angular momentum. For the CT it is to bring the planets inwards from original locations very far from the Sun, i.e. inwards migration with a consequent loss of angular momentum. This is the problem now addressed.

15.7. The Capture Theory and Circumstellar Disks

Although most of the protostar is stretched into a dense filament in the CT process, the CT simulations show that some protostar material moves inwards towards the star. This forms a circumstellar disk, usually of mass 25–50 M_J and extending out to several hundred au. In most simulations the areal density falls off outwards monotonically but quite often it can be of doughnut form with maximum density a few hundred au from the star. In Figure 15.5 the peak areal density is about 350 au from the star.

Simulations of the evolution of planetary orbits, starting with extended eccentric orbits as given by the CT process, were made by Woolfson (2003). The medium was modelled as shown in Figure 15.6. The points were most densely packed close to the star and the masses of individual points were fixed to give the required density distribution. The medium points were in Keplerian orbits and did

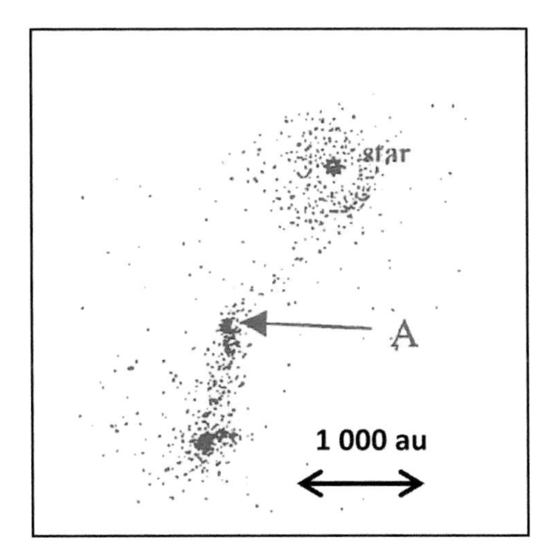

Figure 15.5. A capture-theory simulation showing a strong doughnut-like captured medium (Woolfson, 2003).

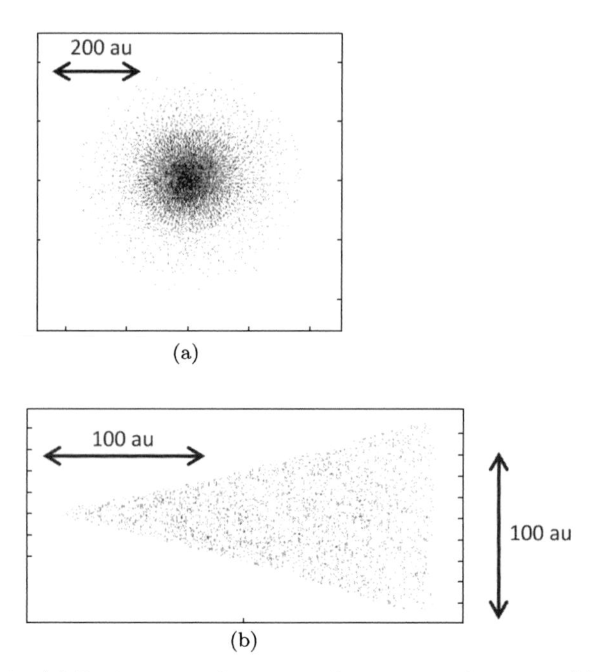

Figure 15.6. (a) Resisting-medium particles seen in plan-view. (b) The distribution perpendicular to the plane (Woolfson, 2003).

not act gravitationally with each other so, in the absence of a planet, the medium was stable. To simulate the decay of circumstellar disks with time the mass of each medium points declines as

$$m(t) = m(0)\exp(-\gamma t), \tag{15.8}$$

where $m(t)$ is the mass of a point at time t.

The areal density for a medium falling off monotonically from a star was modelled as

$$\rho(r) = C\exp(-\alpha r), \tag{15.9a}$$

and for a doughnut form,

$$\rho(r) = C\exp\left\{-\alpha^2\left(r - r_p\right)^2\right\}, \tag{15.9b}$$

where C is adjusted to give the total mass of the medium, and this areal mass is spread perpendicular to the disk uniformly between W

and $-W$ where

$$W = \frac{\pi c}{2\omega}, \tag{15.10}$$

in which c is the sound speed in the medium and ω is the local Keplerian angular velocity.

15.8. The Evolution of Planetary Orbits

To investigate orbital evolution a planet is inserted into the model resisting medium with an initial orbit that a CT simulation might give and its motion followed computationally. Figure 15.7 shows the result of one calculation. The areal density was of the form given by equation (15.9b) with parameters:

Planet mass $= 4\,M_J$; initial semi-major axis $=$ 1,500 au; initial eccentricity $= 0.9$; mean molecular mass for medium $= 2 \times 10^{-27}$ kg; medium temperature $=$ 20 K; total medium mass $=$ 50 M_J; star mass $M_\odot$; with r, the distance from the star in au, $\alpha = 0.007587\,\text{au}^{-1}$; $r_p = 200$ au; with time in years $\gamma = 10^{-6}\,\text{year}^{-1}$; medium represented by 77,408 particles.

The orbit rounded-off after 1.5 million years and after 8 million years decayed to a semi-major axis of about 2.4 au. Some obvious characteristics were revealed by these calculations. For example, a more massive or a longer-lasting medium makes round-off and decay faster and increases the total orbital decay. As the orbit evolves the periastron, q, steadily increases until round-off after which time the circular orbit continues to decay. A less intuitive result is that decay *increases* with the mass of the planet. In Table 15.2, the initial orbital parameters were $(a, e) =$ (1,500 au, 0.90) and the medium of Gaussian form as given by equation (15.9b).

It was impossible to complete some simulations because the semi-major axis became increasingly small, also reducing the required timestep of the calculation. The conclusion from the calculation is that the planet would plunge into the star. However, observed exoplanets have semi-major axes down to 0.015 au and there exists a mechanism that can maintain that order of distance even while the resisting medium is present. It is the mechanism that causes

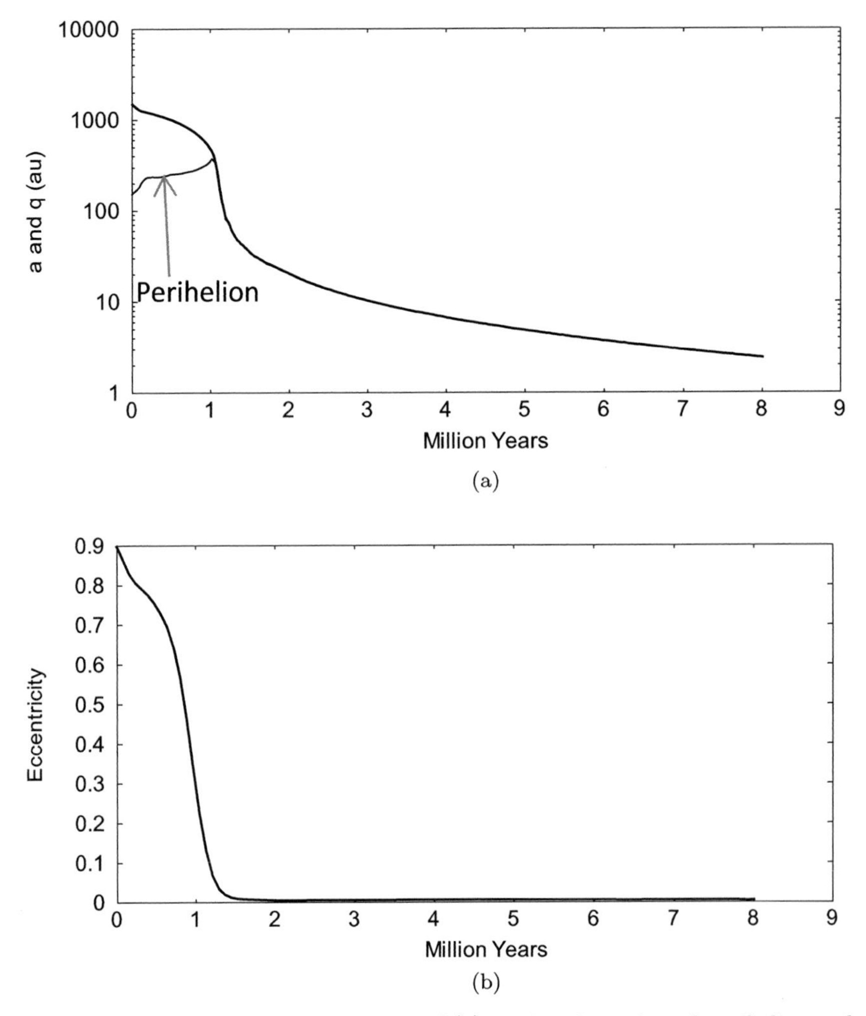

Figure 15.7. The variation with time of (a) semi-major axis and perihelion and (b) eccentricity.

the Moon to retreat slowly from Earth. It depends on the central-body spin period being less than the secondary-body orbital period. The tide raised on the central body by the orbiting body is dragged forward by the central-body spin. The near tidal bulge on the central body gravitationally pulls the orbiter in the direction of its motion thus increasing its orbital angular momentum, which moves

Table 15.2. The variation of the final a and e with protoplanet mass for a given medium.

Mass of planet	Semi-major axis (au)	Eccentricity
M_j	2.980	0.0067
$2M_J$	1.549	0.0065
$3M_J$	1.056	0.0064
$4M_J$	0.787	0.0064
$5M_J$	0.633	0.0062

it outward. In the present context, when the energy gain by the tidal interaction equals the energy loss due to the medium resistance then the orbit stabilizes.

For a low-mass, diffuse or short-lived medium the orbital evolution may not progress very far and the final semi-major axis may be very large. Speculation concerning a possible solar-system Planet 9 with estimated $(a, e) = (700\,\text{au}, 0.6)$ could correspond to a protoplanet produced in an extremely extended heliocentric orbit that only partially decayed and rounded-off.

Orbital evolution occurs throughout a planet's orbit but the general pattern of the evolution can be determined by just considering what happens at periastron and apastron. At periastron the planet moves faster than the medium and slows down. This keeps the periastron constant but reduces the apastron, thus reducing the eccentricity. At apastron the planet moves slower than the medium and speeds up. This keeps the apastron constant but increases the periastron thus reducing the eccentricity. At both extremes the eccentricity decreases and if, as commonly occurs, the periastron effect is stronger then the orbit will also decay. However, there are some exoplanet orbits of high eccentricity, up to 0.97, and we must consider how these could occur.

Young stars generate strong stellar winds that exert an outward force on the medium that opposes the star's gravitational influence. In such a case the medium orbits more slowly, as though the star's mass was reduced. At periastron the planet still moves faster than the medium and so the apastron and eccentricity are reduced. With the slower medium the planet can now move *faster* than the medium at apastron and is slowed down, the effect being that the periastron is reduced and the eccentricity *increased*. For a medium with the doughnut structure shown in Figure 15.5 the orbit would initially decay and round-off but once the apastron reached the vicinity of the peak density the effect there, and going inwards from the peak, would dominate and the eccentricity thereafter increase.

Simulations were made for a planet of mass 5 M_J with initial orbit $(a, e) = (1{,}500\,\text{au}, 0.9)$ and a medium of mass 50 M_J, with the form shown in equation (15.9b). They were run with different diminutions of the effective stellar mass for the medium. For effective stellar mass greater than 0.5 of the true mass the final orbits were circular but for lesser values the final eccentricity steadily increased as the effective stellar mass decreased. Figure 15.8 shows the variation of semi-major axis and eccentricity with time for three runs with different effective masses.

15.9. Exoplanets Around Binary Stars

In Section 13.5, it was mentioned that exoplanets have been detected orbiting one or both stars of a binary system. Since there are as many binary systems as single stars formed in a star-forming environment, interactions between protostars and binary systems should frequently occur. For the CT, with filament distance from the star usually 1,000 au, more or less, a close-binary pair would act like a single gravitational centre of slightly varying strength and direction and the CT process would resemble that for a single star. As a planet's orbit decays in the medium surrounding the stars one of at least three things could happen. It could be left in orbit around both stars, acquire enough energy to escape from the binary system or be captured by one of the stars. For exoplanets around one or both stars there is the possibility that over time the three-body system would become unstable with the planet being expelled to become a free-floating planet (Holman and Wiegert, 1999).

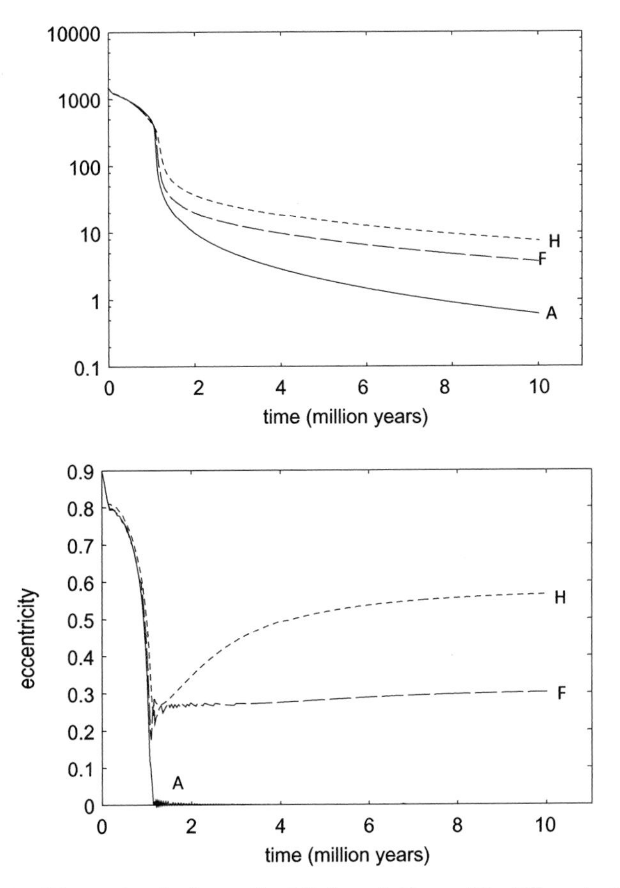

Figure 15.8. Three simulations of orbital evolution with different strengths of stellar wind, two of which give eccentric orbits (Woolfson, 2003).

15.10. Commensurabilities of Planetary Orbits

For stars with many planets, such as the Sun, when the planetary orbits decay they are influenced not just by the star and the resisting medium but also by each other. For the Solar System the ratios of the orbital periods of pairs of the major planets are close to the ratio of small integers, e.g.

$$\frac{\text{Orbital period of Saturn}}{\text{Orbital period of Jupiter}} = \frac{29.46 \text{ years}}{11.86 \text{ years}} = 2.48 \approx \frac{5}{2},$$

and

$$\frac{\text{Orbital period of Neptune}}{\text{Orbital period of Uranus}} = \frac{164.8 \text{ years}}{84.02 \text{ years}} = 1.96 \approx \frac{2}{1}.$$

An explanation of how these orbital commensurabilities became established was given by Melita and Woolfson (1996). In one computational test Jupiter-mass and Saturn-mass bodies were given orbits much further from the Sun than the actual planets but with their orbital periods in the ratio 2:5. The initial eccentricities and inclinations were the same for both orbits — 0.1 and 0.06 radians (3.4°). The model was run three times with different initial relative positions. Both semi-major axes and inclinations fell monotonically but the eccentricities fell at first but then rose again. The final outcome, as seen in Figure 15.9 is that, the ratio of the two periods departed from 2.5 and settled down close to 2.0, actually oscillating about 2.02, although both orbits were still decaying.

The key to understanding this behaviour is that when two bodies are in commensurate orbits the outer body removes energy from

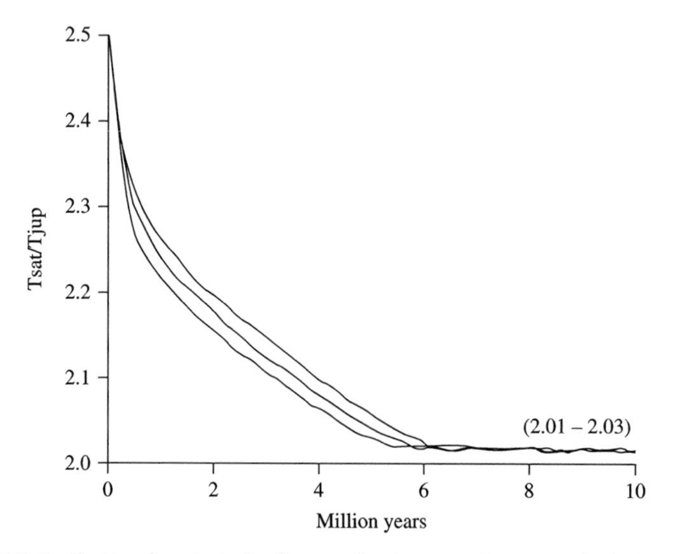

Figure 15.9. Ratio of periods for Saturn:Jupiter starting near the 5:2 resonance. Each curve represents a different initial relative position (Melita and Woolfson, 1996).

the inner one and, conversely, the inner body adds energy to the outer one. At first the eccentricities fall quickly, which reduces the speeds of the planets relative to the resisting medium and the rate of energy dissipation that, in turn, reduces the rate of change of orbital parameters. As a general rule the more massive a planet is, the faster is its orbital decay but, although Saturn is less massive than Jupiter, its orbit decays more rapidly because it is due to Type I migration, which is more effective than the Type II migration affecting Jupiter. When commensurability is established the bodies come closest together repeatedly at the same points of their orbits. This resonance effect amplifies the mutual perturbations of the planets and increases the eccentricity which, in its turn, increases the rate of dissipation and hence the rate of decline in the semi-major axis. However, the increasing dissipation due to the increasing eccentricity of Saturn's orbit is balanced by a gain of energy from Jupiter due to the 2:1 resonance. Thus the rate of change of Saturn's semi-major axis is reduced while that of Jupiter is increased because the effect of Saturn as an exterior body is to add to the loss of energy due to dissipation. The resonance state is now lost and once again Saturn decays more rapidly until resonance is re-established. The result of this is that the ratio of the periods, which had fallen from 5:2 to 2:1, now becomes locked at close to 2:1, with a slight variation, although both orbits still decay.

The numerical experiments were repeated for the Jupiter-Saturn system with different initial eccentricities — nine combinations with each eccentricity having the values 0.1, 0.2 and 0.3 — but always with an initial orbital period ratio 2.5. The results are shown in Figure 15.10. Five combinations end up with a ratio close to 2.0 and four stayed at 2.5. Other trials gave a Neptune–Uranus system with a ratio close to 2.0 starting with the present observed ratio of 1.96 (Figure 15.11).

15.11. The Inclinations of Exoplanet Orbits

In Section 13.4.2, the observed distribution of SOMs was described, showing that although there was a preponderance of small SOMs, there were also many indicating retrograde orbits.

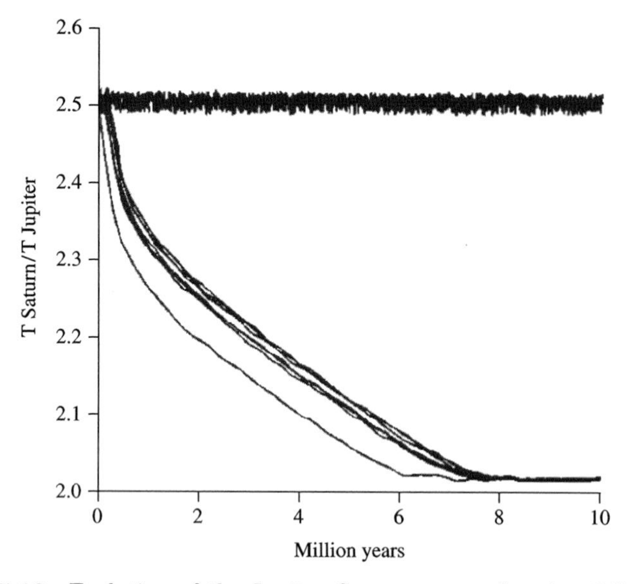

Figure 15.10. Evolution of the Jupiter–Saturn system for nine different pairs of initial eccentricities (Melita and Woolfson, 1996).

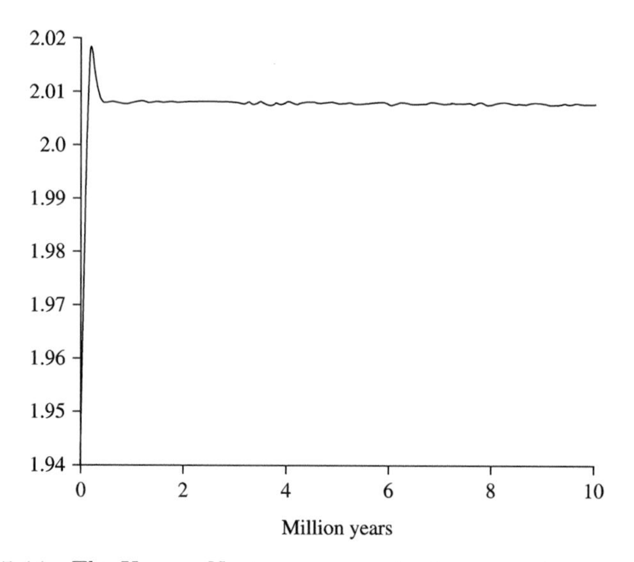

Figure 15.11. The Uranus–Neptune system starting near the present ratio of periods. The system reaches a resonant configuration with a ratio of periods slightly greater than 2 (Melita and Woolfson, 1996).

Because in the CT the spin axis of a star has no relationship to the plane of the star-protostar orbit — which defines the planetary orbits — all SOMs should be possible and, assuming random inclinations, the probability of having an SOM i is proportional to $\sin(i)$, something that Figure 13.8 shows is not true.

Simulations of the CT mechanism not only produce protoplanets and a resisting medium but also adds protostar material to the star, the angular momentum of which pulls the spin axis of the star towards the normal to the exoplanet orbits so reducing the SOM. Some of this material is added by direct transmission from protostar to star but other added material comes from the circumstellar disk, the inner part of which will drift inwards (Lynden-Bell and Pringle, 1974). Comparatively little absorbed material can have a big effect; one-third of a Jupiter mass in orbit at the solar equator has as much angular momentum as the Sun in its spin.

The final SOM depends on the initial value, due to the star–protostar orbit, and the mass of absorbed protostar material. On the basis of this model of explaining the distribution of SOMs the peak in the distribution for small SOMs suggests that there is a high likelihood of substantial absorption of material by the star — possibly a few Jupiter masses.

The CT mechanism leads to a complete range of possible SOMs but with a strong bias towards small values (Woolfson, 2013b).

15.12. Satellites and Angular Momentum

When Galileo saw the large Jupiter satellites through his telescope he interpreted it as a small-scale version of the Solar System. For him, and for most astronomers, it became axiomatic that the process for producing satellites should be a small-scale version of that for producing planets. Thus, planets are produced in circumstellar disks and satellites in circumplanetary disks.

To explore this assumption further we now compare the planetary system and satellite systems in terms of angular momentum distribution by finding the following ratio with respect to a number of primary and secondary pairs of bodies. With 'intrinsic' meaning

Table 15.3. The ratio, S, of the intrinsic angular momentum of the secondary orbit to that of the spin of the central body at its equator.

Central body	Secondary body	Ratio S
Sun	Jupiter	7 800
Sun	Neptune	18 700
Jupiter	Io	8
Jupiter	Callisto	17
Saturn	Titan	11
Uranus	Oberon	21

'per unit mass' this is

$$S = \frac{\text{intrinsic orbital angular momentum of secondary}}{\text{intrinsic angular momentum for material at primary equator}}. \tag{15.11}$$

This quantity is given for various pairs of bodies in Table 15.3. Although for the planetary system the ratio may be affected by magnetic braking of the Sun's spin (Section 14.2.5), the indication is that the orbital angular momentum of satellites does not dominate to the extent that it does for the solar-system planets. This indicates that it is not unreasonable to consider that the processes of forming planets and satellites could be different.

15.13. A Mechanism for Satellite Formation

Figure 15.3 shows that, as a protoplanet collapses, a disk is left behind of mass comparable to, but somewhat less than, that of the collapsing core. Woolfson (2004a) described a process of satellite formation in the disk that exactly parallels that proposed for planet formation for the NT. The steps are those described in Section 14.3.1, except that step (iv), the acquisition of a gaseous envelope, will not occur and the bodies produced in step (ii) will be called *satellitesimals*.

15.13.1. *Dust Settling*

Although the dust particles most easily detected in molecular clouds are of submicron size, there is a distribution of sizes

$$n(D) = KD^{-3.5}, \quad (15.12)$$

where $n(D)$ is the number density of particles with diameter D and K is a constant. Recent work suggests that particle diameters up to $5\,\mu$m are present and Wood *et al.* (2001) detected dust particles up to $50\,\mu$m in size in the dust disk surrounding a T-Tauti star. More recently, millimetre size grains have been detected in some disks. To stay within safe bounds of what will be present in circumplanetary disks, we accept an upper limit of $5\,\mu$m. With this limit almost one-half of the mass of dust is contained in particles between $2\,\mu$m and $5\,\mu$m. The larger particles in the disk move faster towards the mean plane and, as they move, they grow by sweeping up slower-moving smaller particles.

The disk would have a flared structure (Figure 15.6(b)) and an areal density decreasing outwards. Particles closer in have less far to fall but do so in a denser medium that offers more resistance. Settling times at different distances from a planet are shown in Figure 15.12, using theory developed by Weidenschilling, Donn and Meakin (1989) for the following set of parameters.

Planet mass $= 2.0 \times 10^{27}$ kg (approximately the mass of Jupiter)
Disk mass $= 2.0 \times 10^{27}$ kg
Areal density fall by factor e every 2×10^6 km distance from planet
Mean molecular mass of gas $= 4 \times 10^{-27}$ kg
Temperature $= 20$ K
Density of dust grains $= 3 \times 10^3$ kg m^{-3}
Ratio of principle specific heats of gas $= 5/3$

For a disk lifetime of 3 million years there is complete settling beyond a distance of 4×10^6 km. The total mass of Jupiter's large *Galilean satellites* is about 4×10^{23} kg and the total mass beyond 4×10^6 km, both gas and dust, is over 8×10^{26} kg. If just 0.5% of that is deposited

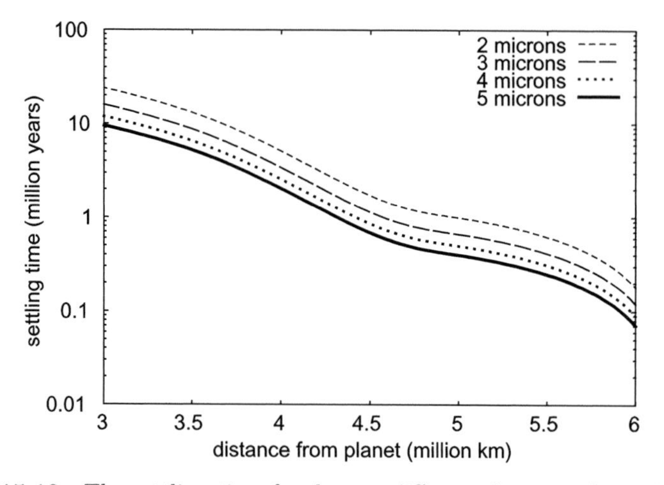

Figure 15.12. The settling time for dust at different distances from the planet and various particle diameters.

dust then the carpet mass would be 4×10^{24} kg — an ample source of solid material to form all of Jupiter's satellites. If larger dust particles were accepted then timescales for settling would be much reduced.

15.13.2. *Formation of Satellitesimals*

The theory for dust-carpet gravitational instability and the formation of planetesimals was given in Section 14.4.2 and this is precisely the theory that can be applied to form satellitesimals. An important difference is that the circumplanetary disk has a much greater average areal density, about 2.5×10^4 kg m^{-2}, than within the circumstellar disk, 1.6×10^3 kg m^{-2}.

The Galilean satellites, Io, Europa, Ganymede and Callisto, have masses $8.93 \times 10^{22}, 4.88 \times 10^{22}, 1.497 \times 10^{23}$ and 1.068×10^{23} kg respectively. At a distance of 4.0×10^6 km, a satellitesimal mass is about 2×10^{22} kg, between about one-half to one-seventh the mass of the Gallilean satellites (Figure 15.13). The next most massive Jovian satellite is Amalthea, closer in than the Galileans and with mass 2×10^{18} kg, within the range of masses of satellitesimals in the figure.

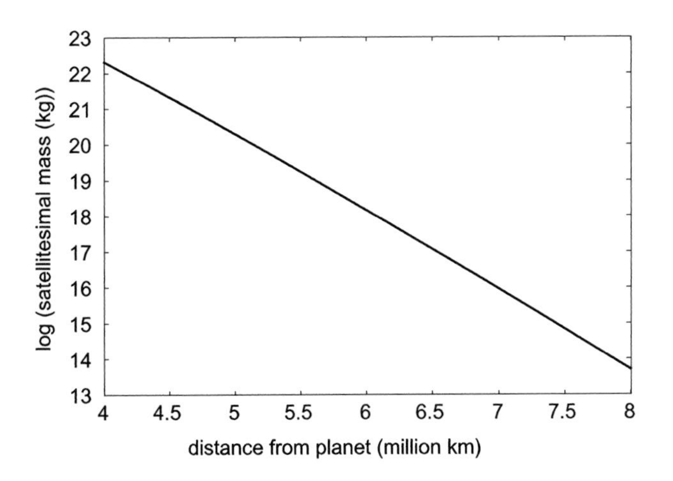

Figure 15.13. Satellitesimal masses at various distances from the planet.

15.13.3. *From Satellitesimals to Satellites*

The expression found by Safronov (1972) for the time to form a planet, equation (14.29), can be applied to satellite formation. We write it as

$$t_{\text{sat}} = \frac{3P_s}{\sigma\,(! + 2\beta)}\left(\frac{4\rho_s}{3\pi^2}\right)^{\frac{2}{3}} M_s^{\frac{1}{3}}, \tag{15.13}$$

where P_s is the Kepler period for the satellite in the region of formation, β is a constant in the range 2 to 5, ρ_s is the satellite density, σ is the mean areal density of satellitesimals and M_s the mass of the satellite. For a satellite of mass 10^{23} kg and density 2.5×10^3 kg m^{-3}, out to a distance 4.0×10^6 km all formation times are less than 500,000 years.

For the model disk being considered, a dust carpet only formed beyond a distance of 4.0×10^6 km within 3 million years but the orbital radius of Callisto, the outermost Galilean satellite, is 1.88×10^6 km. The answer to this apparent discrepancy is that protosatellite orbits are decaying by the Type I migration mechanism as they grow and will continue to decay for the lifetime of the disk. Figure 15.14 shows the decay of a partially formed satellite, with constant mass 10^{22} kg, moving in the disk that gave Figure 15.12, but decaying so

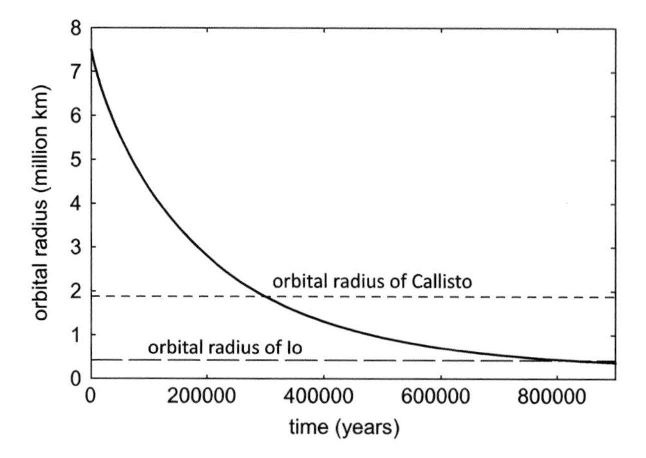

Figure 15.14. The decay of the orbit of a satellitesimal of mass 10^{22} kg (After Woolfson, 2004a).

that the density falls everywhere by a factor of e every million years. Starting with a radius of 7.5×10^6 km the orbit decays to the distance of Callisto, the outermost Galilean satellite, in about 3×10^5 years and to the orbit of Io, the innermost one, in about 8×10^5 years. These short decay times are due to the relatively high areal density of the protoplanetary disk. The orbit is close to circular for the whole of the decay period.

There are many commensurabilities found for the periods of pairs of satellite orbits of Jupiter and Saturn. These are due to coupling between satellites as their orbits decayed, as described in Section 15.10.

15.14. The Problem of the Terrestrial Planets

In the CT, protoplanets form because the mass of a condensation within the protostar filament, mostly gas, is greater than the Jeans critical mass. The simulations show that the protoplanets initially move towards apastron and by the time they approach periastron they have become compact bodies capable of resisting tidal disruption; gaseous exoplanets, so-called *hot Jupiters*, are observed at distances down to 0.015 au from their stars. It seems that the CT can only give major planets, raising the question of how terrestrial planets formed.

Dormand and Woolfson (1977) first proposed the idea of a planetary collision to explain terrestrial planets but the knowledge and computer technology of the time did not enable a realistic model to be created. They pointed out and demonstrated that when the planetary orbits were evolving, the mass of the circumsolar disk gave a non-central force on them, causing orbital precession. Differential rates of precession meant that slightly inclined pairs of orbits occasionally intersected during the evolutionary period. Dormand and Woolfson postulated an initial system of six planets, the present four major planets plus two others, and found that the probability that some pair would collide before they all rounded-off was of order 0.1 — small but not negligible. A NASA Spitzer Space Telescope observation in August 2009 gave evidence of a planetary collision in the vicinity of the young star HD172555 (age 12 million years) within the last few thousand years. In the light of new knowledge and the availability of suitable computational tools the planetary-collision hypothesis has been revisited (Woolfson, 2013b).

15.15. Deuterium in the Colliding Planets

The distribution of deuterium — the stable isotope of hydrogen represented by the symbol D — within the colliding planets plays an important role in the collision process. The cosmic *D/H* ratio is about 2×10^{-5} but in star-forming clouds, while the *overall D/H* ratio is similar to the cosmic ratio, it is extremely non-uniform. The *D/H* ratio in some molecular species in these clouds, and in low-mass protostars formed within them, is very high (Roueff *et al.*, 2000: Loinard *et al.*, 2001; Loinard *et al.*, 2002; Parise *et al.*, 2002). The average over all the molecular species is estimated as $D/H > 0.01$. This concentration of deuterium is due to the phenomenon of *grain-surface chemistry.* A deuterium atom falling on the surface of an icy grain exchanges places with a hydrogen atom in a molecule because this lowers the energy of the molecule and hence increases its stability. Over a long period this process concentrates the deuterium in ice molecules which form either icy grains or ice coatings on silicate or iron grains. In cold clouds, in addition to the common hydrogen-containing molecules — water, ammonia and methane — more complex molecules are present in considerable quantities. The ratio of methanol to water, CH_3OH/H_2O, is in the range 0.1 to 0.5.

A protoplanet substantially collapses in about 10^4 years (Figure 15.3) and thereafter evolves slowly to its final state. During the early slow free-fall stage of collapse, solid grains migrate towards the centre. Eventually an iron core with a silicate mantle forms, surrounded by a shell of vaporized hydrogen-containing molecules with a high *D/H* ratio. Over a long period of time the excess deuterium migrates outwards to increase the *D/H* ratio in the gaseous envelope. The importance of deuterium is that D–D nuclear reactions take place at low temperatures, just a few million K, whereas most nuclear reactions require temperatures greater than 10^8 K.

15.16. The Planetary Collision; Earth and Venus

Following the proposal of Dormand and Woolfson, it was postulated that the initial Solar System contained no terrestrial planets but two extra major planets, Enyo of mass 1.9 M_J and Bellona of mass 2.5 M_J (Woolfson, 2013b). These extra planets were modelled in four layers based on incomplete settling of material by density — an iron core with some silicate, a silicate mantle, with some iron, a deuterium-rich gaseous shell with some silicate and a hydrogen–helium atmosphere. The overall composition is given in Table 15.4. Point-mass models for the SPH simulation were formulated, as described by Woolfson (2007), in which the density of SPH points was highest at the centre, the region of greatest interest for this simulation. Each planet was represented by 4,921 SPH points. A Tillotson equation-of-state (Tillotson, 1962) was used for the inner three regions and a modified gas law for the atmosphere, which accommodated the high pressure regions where the perfect gas law breaks down. Starting with the temperature and density at the centre given by Table 15.4, the equations for gravitational and pressure equilibrium were integrated outwards with the criteria that the density was discontinuous at boundaries

Table 15.4. The characteristics of the colliding planets.

Planet	Bellona	Enyo
Mass ($M_\oplus$)	799	598
Radius (10^4 km)	9.152	8.647
Central density (10^3 kg m^{-3})	176.5	146.5
Central temperature (10^3 K)	85	74
Mass of iron ($M_\oplus$)	3.00	2.50
Mass of silicate ($M_\oplus$)	12.00	10.00
Mass of ice ($M_\oplus$)	6.00	5.00

between regions but that temperature was continuous. The planet's outer boundary was taken when the temperature fell to 100 K.

The planets moved towards each other on parallel paths with an offset of 7×10^4 km and had a contact speed of 90 km s^{-1}. When the shock front reached the Enyo deuterium-rich region the temperature gave a high rate of D–D nuclear reactions. Although at the time the enrichment of deuterium in grains in star-forming regions had not been observed, Holden and Woolfson (1995) examined in detail the nuclear reactions that would occur with a mixture of deuterium-rich ices and silicates. Once the temperature reached $\sim 5 \times 10^8$ K the deuterium was exhausted and heat generation by other thermonuclear reactions was at a lower rate. In the present simulation, when an ice SPH particle reached a temperature of 3×10^6 K (when the rate of D–D nuclear reactions was high) the temperature was immediately raised to 4×10^8 K, lower than the Holden and Woolfson results indicated. This simplified the incorporation of the nuclear reactions without exaggerating their effect. The locations in which nuclear reactions took place quickly spread to other regions of *D/H* enhancement.

Figure 15.15 shows the progress of the simulation. By frame (c) D–D nuclear reactions had occurred as seen by the rapid outward motion of material. This expansion became greater for successive frames but parts of the cores remained compact and steadily moved apart.

The approach speed of the planets when far apart, 42.9 km s^{-1}, suggests a collision in the terrestrial region. It is proposed that the

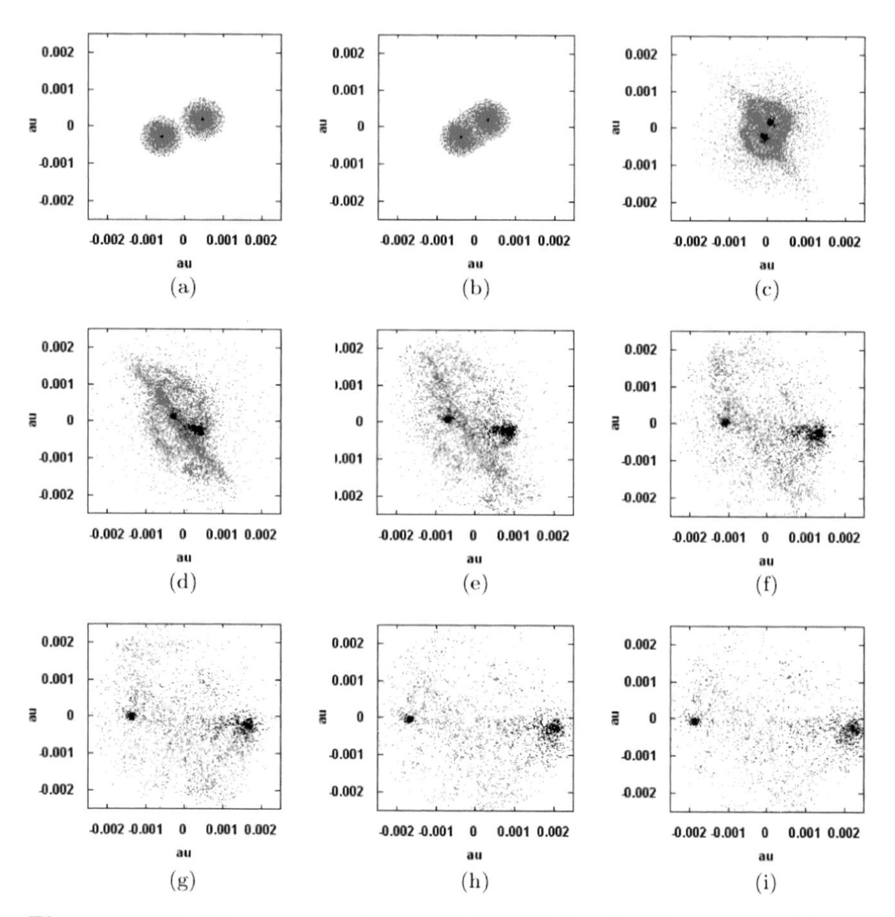

Figure 15.15. The progress of the collision. (a) $t = 0$, (b) $t = 590\,\text{s}$, (c) $t = 1{,}326\,\text{s}$, (d) $t = 2{,}505\,\text{s}$, (e) $t = 3{,}917\,\text{s}$, (f) $t = 5{,}336\,\text{s}$, (g) $t = 6{,}415\,\text{s}$, (h) $t = 7{,}597\,\text{s}$, (i) $t = 8{,}609\,\text{s}$ (Woolfson, 2013b).

residual cores seen in Figure 15.15 formed the Earth and Venus, from the Bellona and Enyo residual cores, respectively. Their estimated masses from the simulation, 2.4 $M_{\oplus}$ and 1.5 $M_{\oplus}$, are both too large, but the model indicates how the larger terrestrial planets could have originated.

The orbits of Earth and Venus, formed in this way, while a resisting medium was still in place, would have evolved by the Type I migration mechanism within the terrestrial region.

15.17. The Moon

The present solar-system major planets all have large families of satellites. Most of them are *regular satellites*, formed by the process described in Section 15.13. They tend to have close-to-circular orbits near the plane of the planetary equator and to be within the inner part of the satellite system. Satellites not satisfying these conditions, *irregular satellites*, are normally taken as captured bodies.

Bellona and Enyo would have had many satellites and, because of their masses, some of the satellites would almost certainly have been more massive than Ganymede, the most massive solar-system satellite. Following the collision, the possible outcomes for a particular satellite are limited to the following:

(I) The satellite could be retained by one or other of the residual cores.
(II) The satellite could end up in a heliocentric orbit.
(III) The satellite could escape from the Solar System.
(IV) The satellite could be disrupted by debris from the collision.

Test calculations show that the most likely outcomes are (II) and (III). Here it is proposed that the Moon was a satellite left in orbit around the Bellona residual core — now Earth. In both mass and density it is intermediate between the Galilean satellites Io and Europa, supporting the view that it was formed by the normal process described in Section 15.13.

In 1959, the Soviet Union Luna 3 spacecraft sent back low-quality images from the Moon that were good enough to show that, although the nearside is dominated by large maria, the far side is predominantly heavily-cratered highlands. Lunar satellite altimeter measurements revealed that the far side has large impact basins, so the Moon had been bombarded uniformly by large projectiles, but the far-side basins had not filled with magma from below to give maria. It was suggested that this was due to a difference of crustal thickness on the two sides, which was confirmed when seismometers were left on the Moon by Apollo astronauts. While crustal thickness

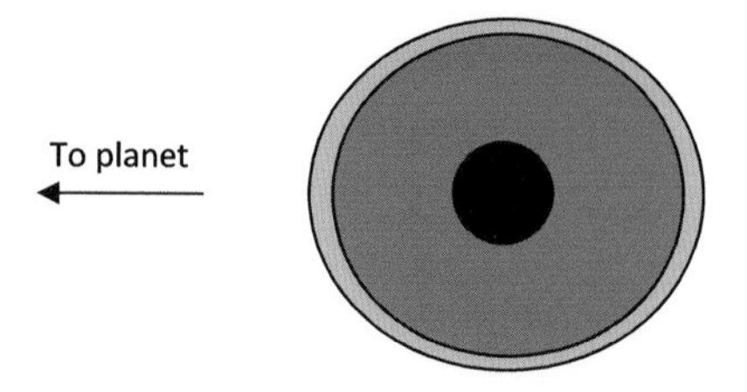

Figure 15.16. The initial structure of a satellite formed in synchronous orbit around a planet showing core (black), mantle (dark grey) and crust (light grey). The satellite distortion and thickness of the crust are exaggerated.

varied from place to place, the average thickness on the far side is about 12 km greater than that on the nearside.

An early fluid or plastic satellite in synchronous orbit around a planet should have a *thicker* low-density crust on the nearside (Figure 15.16). The Moon formed around either Bellona or Enyo would have acquired a figure and distribution of material that would have ensured that when it orbited the residual core it should eventually have presented the same thicker-crust face to the Earth.

Due to near-surface convection, driven by solidified surface material sinking in the less dense liquid material below, the lunar crust would have solidified to some depth in the time between lunar formation and the planetary collision — probably of order 1 million years. The nearside of the Moon, facing the collision, was bombarded by debris travelling at about 100 km s^{-1}. Sharing the debris energy with lunar surface material, given that the escape speed from the Moon is 2.4 km s^{-1}, would have led to massive abrasion of the nearside and it has been estimated that up to 50 km thickness of surface material could have been removed in this way (Woolfson, 2013b). This estimate requires that the arriving debris removed up to about eight times its own mass that, since the energy of the arriving debris has 1,600 times the intrinsic energy of escape from the Moon, is quite feasible.

The current most-supported model for the formation of the Earth-Moon system proposes that, shortly after solar system formation, a Mars mass body (Theia) struck the Earth obliquely. From the debris of the collision the Moon accreted in orbit around the Earth.

This scenario has been realistically simulated by Benz, Slattery and Cameron (1986). Based on the difference of the ages of highland rocks and meteorites, as found by radioactive dating, Benz *et al.* estimated that the Theia event occurred 30–100 million years after solar-system formation. Radioactive dating gives the time at which radioactive material solidified, so that the daughter products of the radioactivity are retained in the vicinity of the parent product. Maria, plus the unfilled basins on the far side, cover about one-third of the Moon's surface At their centres the magma depth can be 5–6 km (Head, 1976), indicating an original excavation depth of order 7–8 km. Making an extremely conservative assumption that the average depth of maria excavation was only 1 km and that only 10% of the ejecta was retained and spread uniformly over the lunar surface, the average depth of cover in the highlands would be 30 m — probably a gross underestimate. The highland rocks from the Apollo 16 mission probably come from the nearby Mare Nectaris, the solidification ages of which would depend on their excavation depth. Any conclusions about the time of formation of the Moon based on dating highland surface rocks cannot possibly be valid.

15.18. Mars and Mercury

Mars and Mercury, with about four times and twice the mass of Ganymede respectively, are proposed as escaped satellites. Their relatively-high orbital eccentricities, 0.093 and 0.206, may indicate that the orbits of these small bodies evolved slowly and that their orbital evolution terminated before round-off because of loss of the circumsolar disk.

Like the Moon, Mars has hemispherical asymmetry with magma-covered northern plains and heavily-cratered southern highlands with one large deep depression, the Hellas Basin (Figure 15.17). A 2 km high scarp separates the two hemispheres. Heavy abrasion of one face could have removed most solid crustal material on the exposed side and volcanism would then have produced the northern magma plain. The Hellas Basin represents the effect of an exceptionally energetic projectile that penetrated the crust of the southern highlands.

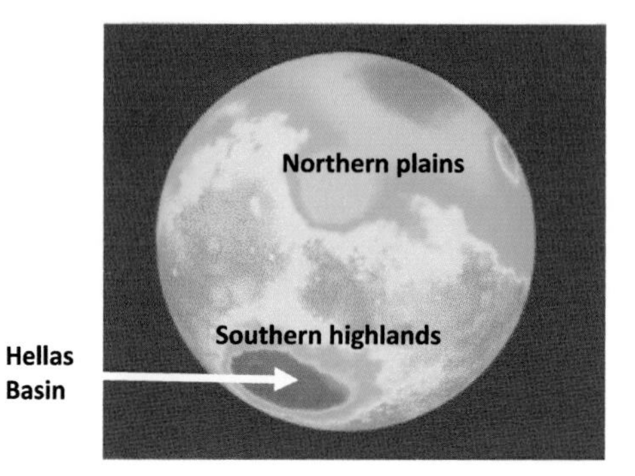

Figure 15.17. The topography of the Martian surface (NASA/JPL/Main Space Science Systems).

Due to tidal coupling to the Earth, the Moon's spin axis is contained within its plane of asymmetry. For Mars, not linked to another body, the plane of asymmetry makes an angle of 35° with the spin axis. Mars would have had a molten mantle over which the lithosphere could move — something akin to continental drift. A theorem by Lamy and Burns (1972) states that a spinning body with internal energy dissipation eventually settles down with its spin axis along the principal axis of maximum moment of inertia, a process known as *polar wander*. McConnell and Woolfson (1983) modelled the surface features of Mars either as positive features — raised regions in the highlands such as the Tharsis Uplift, Argyre Plain, Elysium Plain and Olympus Mons — or as negative features such as the northern plains and the Hellas Basin. They calculated that the principal axis of maximum moment of inertia was 11.9° from the spin axis. The probability of being this close just by chance is about 0.02 and the discrepancy might be due to the crudeness in modelling surface features, or, perhaps, that polar wander was incomplete before the underlying mantle became too rigid to maintain it.

The uncompressed density of Mercury is higher than that of any other planet. Its iron core is similar in size to that of Mars and it has been suggested that it was once similar to Mars but had much of its mantle stripped away by a collision with another body. If Mercury had been a close satellite, or in the plane of the collision

Figure 15.18. The Caloris Basin (NASA).

interface where the debris density was very high (Figure 15.15), then a large proportion of its mantle that faced the collision would have been stripped away. When it reorganized itself into a near-spherical body the motion of surface material, flowing round the surface and meeting in one small region, could have resulted in an overshoot with the following collapse giving the 'bullseye' feature, the Caloris basin (Figure 15.18) and the diametrically opposite Chaotic Terrain, which regions are sub-solar at alternate perihelion passages (Woolfson, 2011b). The Caloris basin–Chaotic terrain axis would have been tidally locked in the direction of its parent planet giving a distribution of matter within Mercury that gives its present relationship with the Sun. This is consistent with the Caloris Basin having been the region facing the collision.

15.19. The Neptune–Pluto–Triton System

Pluto, once considered a member of the planetary family but now categorized as a *dwarf planet*, has an orbit of eccentricity 0.249 and inclination 17° that passes just within the orbit of Neptune, although, due to a 3:2 commensurability of their orbital periods, these bodies

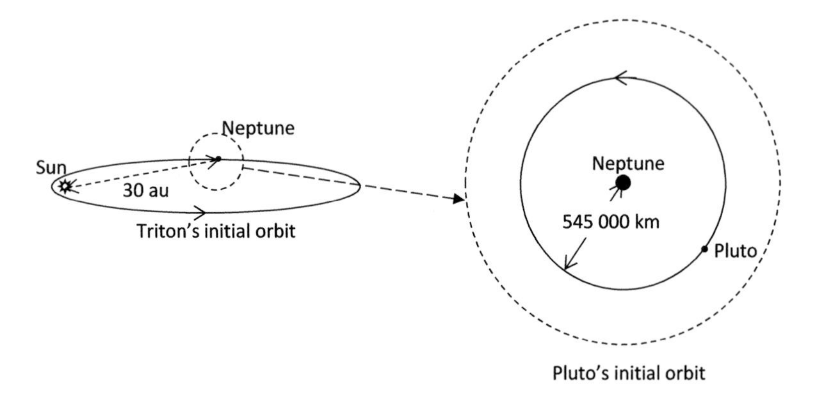

Figure 15.19. The initial orbits of Triton and Pluto before the collision.

never approach closely. Triton, the seventh largest satellite in the Solar System, is in a retrograde orbit around Neptune, which rules it out as a regular satellite. The relationship between these bodies has been explained as another outcome of the planetary collision (Woolfson, 1999).

The scenario is that Triton was a satellite of a colliding planet released into an extended heliocentric orbit taking it beyond Neptune. Pluto was the largest regular satellite of Neptune with mass about two-thirds that of Triton. A computer simulation was made of a collision involving Triton and Pluto. The starting point is illustrated in Figure 15.19. Triton was in a direct heliocentric orbit with perihelion 2.6 au and aphelion 55.6 au and Pluto was in a direct circular orbit, of radius 545,000 km, around Neptune. The before-and-after collision situations are shown in Figure 15.20. Triton, moving inwards, struck Pluto a glancing blow that ejected it into a heliocentric orbit with $(a, e) = (39.5\,\text{au}, 0.253)$, very similar to its present orbit. The glancing collision set Pluto into retrograde spin and a large part of it was sheared off to form its large satellite Charon, with more than one-half the diameter of Pluto, in a retrograde orbit — a process similar to the Benz *et al.* Moon-formation process. Collision fragments formed Pluto's smaller satellites. The four known smaller satellites of Pluto all have retrograde orbits, near-commensurate with that of Charon.

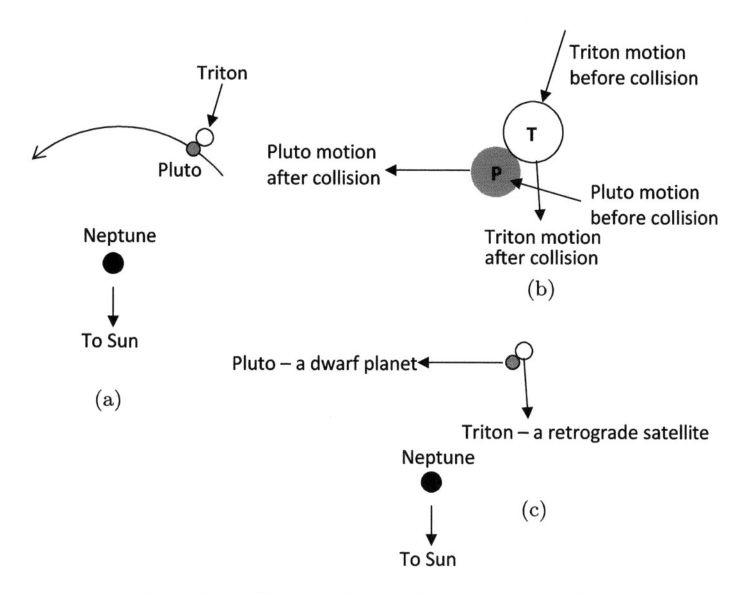

Figure 15.20. The Triton-Pluto collision. (a) triton, travelling towards the Sun strikes Pluto. (b) motions before and after the collision. (c) the final outcome.

Triton lost energy in the collision and was captured by Neptune into a *retrograde* orbit with $(a, e) = (436{,}000\,\text{km}, 0.88)$. Tidal effects give rapid round-off and decay of retrograde satellite orbits (McCord, 1966), so giving Triton's present orbit with $(a, e) = (355{,}000\,\text{au}, 0.000)$.

15.20. Asteroids and Comets

The residual cores, forming the Earth and Venus, account for only a small part of the inventory of iron and silicate in the original planets. The debris, thrown out in all directions, as seen in Figure 15.15, must contain the remainder of this material plus virtually all the icy materials. Some ice may be incorporated in the residual cores and manifest itself as the water and methane now present on and in the Earth. If the intrinsic energy of debris is positive then it will leave the Solar System. Figure 15.21 shows separately for iron, silicate and ice debris the mass in Earth units per unit intrinsic energy, expressed in GJ/kg.

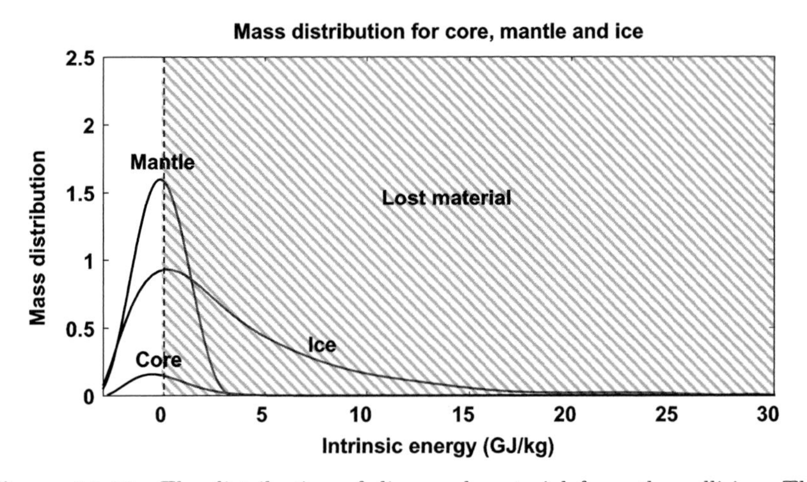

Figure 15.21. The distribution of dispersed material from the collision. The graphs show the distribution of mass in Earth units per unit intrinsic energy (GJ/kg) (Woolfson, 2013b).

More than one-half of the core material is retained and a somewhat smaller fraction of mantle material but most of the ice is expelled from the system. The further out from the planetary centre that material originated, the greater is its volatile content and the further out it would tend to end up in the Solar System. Observations, based on comparing meteorite and asteroid reflection spectra confirms this statistical relationship between volatility and distance from the Sun.

The eccentricities and inclinations of the retained material show some interesting trends, as seen in Figure 15.22. Nearly all the iron and mantle debris orbits are prograde but a considerable proportion of the ice orbits are retrograde.

Comets are associated with two regions of the Solar System — the Kuiper Belt (KB), beginning just beyond Neptune's orbit, and the Oort Cloud (OC), tens of thousands of au from the Sun. It has been suggested that OC comets must have an external origin because they have D/H ratios 20 times the cosmic value but the high ice D/H ratios of the colliding planets weakens that argument. Bailey (1983) proposed that there are comets, between the OC and the KB, which are drawn outwards to replenish the OC when it is depleted by a

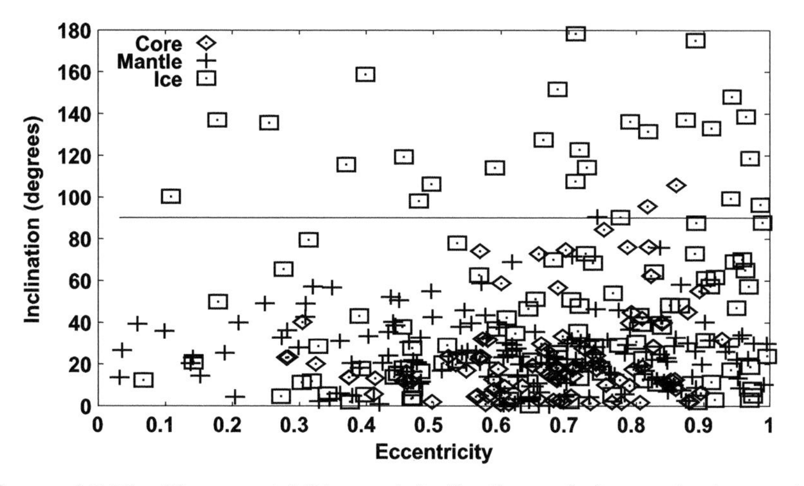

Figure 15.22. The eccentricities and inclinations of the retained material (Woolfson, 2013b).

close stellar passage or when the Solar System has passed through a Giant Molecular Cloud. The distributions for silicate and ice in Figure 15.21, which stretch from the inner Solar System continuously to the region of the OC, are consistent with the Bailey model. The KB bodies, perturbed by Neptune to give short-period comets, form the inner boundary of this distribution. So-called *new comets* are bodies perturbed inwards by external sources from the outer part of the distribution — the OC. Since there are no major perturbation sources for comets between the KB and OC their presence is not detected.

Initially the debris orbits repeatedly pass through the inner Solar System and unless an orbit evolves to give a perihelion outside Neptune's orbit it will almost inevitably either be eventually swept up by a major planet or be ejected from the Solar System.

Asteroid bodies, considered as debris, would have interacted with each other and with planets. Some would have attained safe orbits, such as those in the Asteroid Belt. The total mass of the surviving asteroids is about 4% that of the Moon, a small fraction of the original debris.

Some volatile debris could have interacted with major planets in evolving orbits at hundreds of au from the Sun and be swung into orbits well outside the present planetary region. With aphelion and aphelion represented by Q and q respectively, an interaction of a comet with $(Q, q)) = (110\,\text{au}, 0.5\,\text{au})$ with a Jupiter-mass planet with $(Q,\ q) = (100\,\text{au},\ 10\,\text{au})$ with closest approach $1.84 \times 10^6\,\text{km}$ gave a final orbit for the comet $(Q, q) = (109.9\,\text{au},\ 42.1\,\text{au})$, which would place the comet well within the KB. However, it is unlikely that many fragments were affected in this way.

Small bodies such as comets do not have sufficient mass to have an appreciable gravitational effect on the resisting medium and the resistance is primarily due to the ram pressure they experience, like the pressure on a yacht sail due to the wind. For a spherical comet the force experienced will be

$$\mathbf{F} = \pi \rho a^2 V \mathbf{V}, \tag{15.14}$$

where ρ is the local medium density, a the comet radius and $\mathbf{V}$ the velocity of the medium relative to the comet. The effect of such a force has been found for a comet of mass $7 \times 10^{12}\,\text{kg}$, of density $500\,\text{kg m}^{-3}$ (published estimates are between 100 and $1{,}000\,\text{kg m}^{-3}$), with original perihelion 0.5 au and a range of original semi-major axes. The medium had a total mass of $40\,M_J$ with an annular distribution of density, similar to that seen in Figure 15.5, given by

$$\rho(r, z) = C \exp\left(-\frac{(r-d)^2}{2\sigma_r^2}\right) \exp\left(-\frac{hr^2}{s^2}\frac{z^2}{\sigma_z^2}\right), \tag{15.15}$$

where $d = 100\,\text{au}$, $\sigma_r = 30\,\text{au}$, $h = 10$, $s = 20\,\text{au}$, $\sigma_s = 30\,\text{au}$ and r and z (distance from the mean plane) are expressed in au. The constant C is determined from the total mass of the disk. The results are shown in Figure 15.23. It will be seen from the figure that all orbits with original semi-major axes greater than about 60 au end up with perihelia beyond Neptune's orbit and within the KB region. Their aphelia stretch out to several hundred and even thousands of au and represent Bailey's inner cloud of comets that form a replenishment reservoir for the OC.

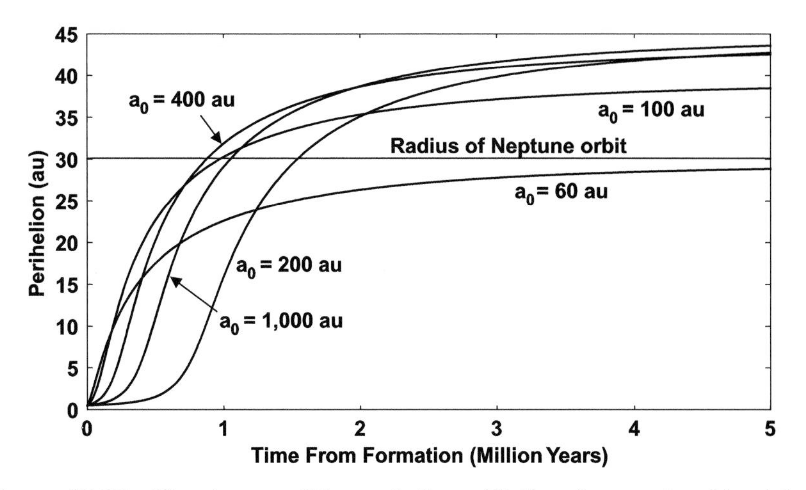

Figure 15.23. The changes of the perihelion with time for comets with original perihelia 0.5 au and various initial semi-major axes, a_0 (Woolfson, 2013b).

15.21. Dwarf Planets

The Moon, Mars, Mercury and Triton have been identified as large onetime satellites of the colliding planets. A class of bodies known as dwarf planets, six in number — Ceres, Pluto, Eris, Makemake, Haumea and V774104 — all have masses within the range of solar-system satellites and we also identify them as ex-satellites.

The orbits of bodies of satellite mass in heliocentric orbits are modified by the Type 1 migration process (Section 14.4.1), so slowly that it is unlikely that orbital evolution will progress to the round-off stage within the lifetime of the circumstellar disk. Figure 15.7(a) shows that the perihelion increases up to the time of round-off. A satellite with an orbit that evolved so that the final perihelion was within the KB would survive over the longer term. Others would eventually be swept up by a major planet or expelled from the Solar System — except for Ceres that orbits within the Asteroid Belt.

There would certainly have been many other ex-satellites, some possibly more massive than the presently-known dwarf planets. They could have escaped from the Solar System directly as a result of the collision, could have been absorbed by major planets or be in outer regions of the KB and be, as yet, undetected. One or more of them may have reached the OC. There is a tendency for a group of new comets to come from similar directions with similar orbital parameters, which may be due to the presence of major perturbing bodies within the OC.

If dwarf planets were redefined as ex-satellites large enough for their self-gravity to mould them into hydrostatic equilibrium and in heliocentric orbits then, according to this model, at present there would be eight — Ceres, Pluto, Eris, Makemake, Haumea, V774104, Mars and Mercury.

15.22. The Ice Giants

The ice giants, Uranus and Neptune, are characterized by three properties:

(i) Low masses compared with the gas giants, due to much smaller gaseous atmospheres.
(ii) Large spin-axis tilts – 98.7° for Uranus, although that for Neptune, 28.3°, is similar to that of Saturn, 26.7°.
(iii) Atmospheric *D/H* ratios higher than both the cosmic average and those of the gas-giants.

There are several possible explanations for each of these properties individually.

It has often been suggested that the large axial tilt of Uranus was due to a tangential collision by a large body; if the axial tilt was originally zero then it would require the colliding body to have several Earth masses and an impact speed of about 100 km s^{-1}.

A clue as to the history of Uranus and Neptune is given by the *D/H* values in their atmospheres. The values for the two pairs of giant planets are;

Jupiter, $2.25 \pm 0.35 \times 10^{-5}$ and Saturn, $1.70^{+0.75}_{-0.45} \times 10^{-5}$ (Lellouch *et al.*, 2001)

Uranus, $4.4 \pm 0.4 \times 10^{-5}$ and Neptune, $4.1 \pm 0.4 \times 10^{-5}$ (Feuchtgruber *et al.*, 2013).

If the star-forming cloud had an overall *D/H* ratio of 2×10^{-5}, if 75% by mass of the gas was hydrogen and that hydrogen in ice, with a *D/H* ratio of 0.015, formed 0.1% of the mass of the cloud then, on this basis, the *D/H* ratio in the gas was 6.7×10^{-6}. The ratio of hydrogen in gas form to that contained in ices is 750:1 and any planet initially with a lesser ratio will have a final *D/H* ratio greater than the universal value once the deuterium has diffused from the ice into the atmosphere.

For the CT the original spin-axis tilts should not have been large. There is a scenario, consistent with the CT, which explains all three of the characteristics of the ice giants in terms of a single event for each of them. It proposes that the ice giants originally had more extensive atmospheres that were partially stripped off in a close oblique collision with a massive planet — taken as Bellona in our simulation — soon after their formation so that their ices still retained the original large *D/H* ratio. This collision also gave a loss of mass and a large final spin-axis tilt SPH models of a proto-ice-giant (PIG) and Bellona were formulated, as described by Woolfson (2007), based on a planet in four distinct layers. For recently-formed planets segregation by density would not be complete so the four layers had compositions as specified below.

core, consisting of 40% iron with the remainder silicate,
mantle, consisting of 85% silicate with the remainder iron,
silicate + ice layer with 10% of the mass (~25% by volume) as ice,
atmosphere, consisting of hydrogen and helium.

The Tillotson equations of state were used for iron, silicate and ices and previously-published models of Jupiter and Saturn (Stevenson and Salpeter, 1976) were used to choose a best model for an atmosphere of the form

$$p = \frac{\rho k T}{\mu}(1 + c\rho), \tag{15.16}$$

with $c = 0.08$. The PIG had mass $54.58\,M_\oplus$ and radius 5.049×10^4 km. The compositions, by mass, of the inner three layers are given

Table 15.5. The inner composition of the colliding planets.

Planet	**Core ($M_\oplus$)**	**Mantle ($M_\oplus$)**	**Ice ($M_\oplus$)**
Bellona	3.00	12.00	6.00
Proto-ice-giant	1.25	5.00	2.50

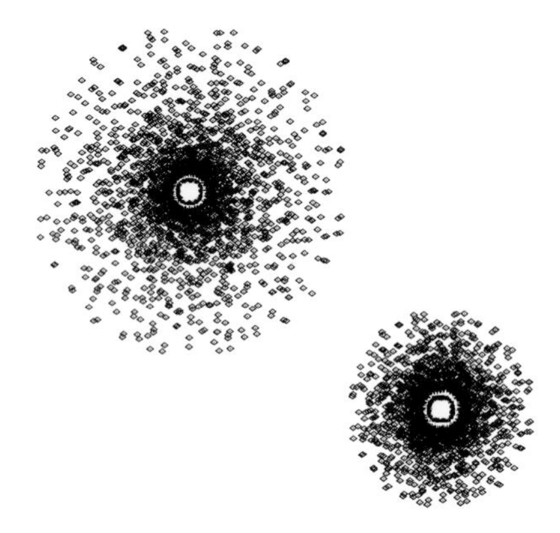

Figure 15.24. The initial configuration for the grazing collision.

in Table 15.5. The density of points, 3,319 in total for each body, is highest at the centre of each body, the region of greatest interest, and falls off towards the boundary. An initial configuration of the two planets is shown in Figure 15.24; the motion of the PIG along the $x-$direction is displaced relative to Bellona along the $y-$direction by a distance 10^5 km. Both bodies initially have spin axes in the $z-$direction with a spin period 10 h, similar to the periods of Jupiter and Saturn.

The progress of the oblique collision can be followed in Figure 15.25 which shows the arrangement of the two bodies at four times relative to zero time for Figure 15.24. The PIG is considerably disrupted and distorted, as is seen in frames (a) and (b), but quickly reassembles itself into a spherical form.

The mass and radius of the residual ice giant (RIG), as seen in Figure 15.25(d) are given in Table 15.6. The RIG retains all the core, mantle and ice material of the PIG so the final mass of

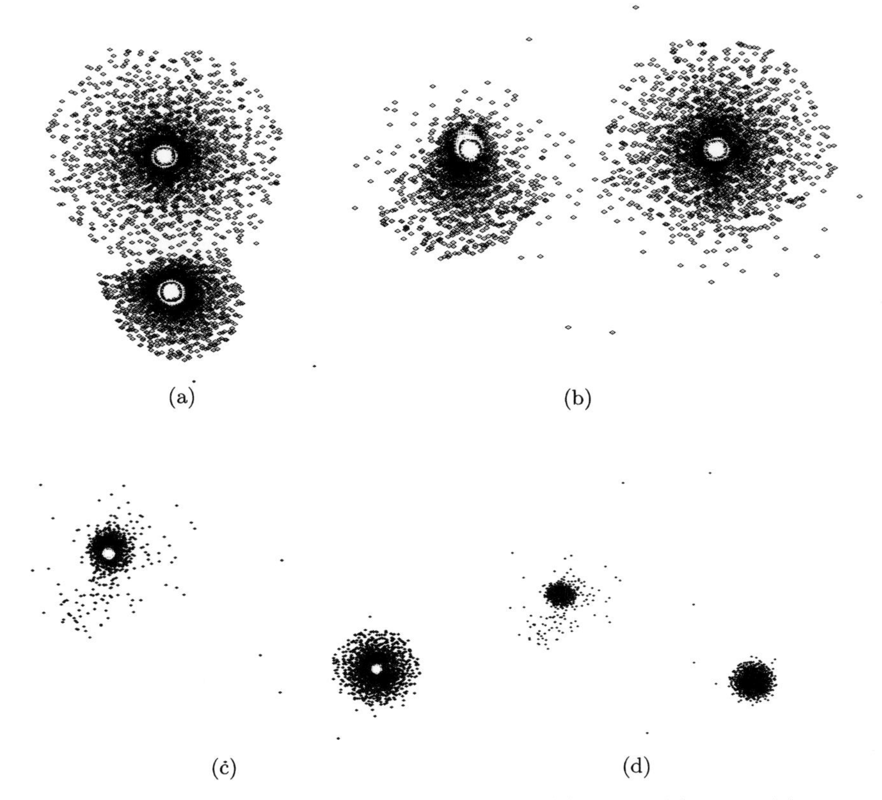

Figure 15.25. The progress of the collision after (a) 0.56 h, (b) 1.33 h, (c) 3.31 h, (d) 4.18 h. The scales vary but the Bellona condensation remains at an approximately constant size.

Table 15.6. Characteristics of Uranus, Neptune and the RIG.

	Uranus	**Neptune**	**RIG**
Mass ($M_\oplus$)	14.54	17.15	18.93
Radius (10^4 km)	2.556	2.476	2.543
Mean density (10^3 kg m^{-3})	1.27	1.64	1.64
Spin-axis tilt (°)	97.8	28.3	68.7
D/H (10^{-5})	4.4	4.1	4.91
Moment-of-inertia factor	0.225	unknown	0.220
Spin period (hours)	7.24	16.11	17

the atmosphere is 10.18 $M_\oplus$, a fraction 0.222 that of the original atmospheric mass of 45.88 $M_\oplus$. The mass of hydrogen in the final atmosphere is $0.75 \times 10.18\,M_\oplus = 7.635\,M_\oplus$. It is now necessary to estimate how much hydrogen is contained in the 2.5 $M_\oplus$ of material described as ice, which is really ice-impregnated silicate material. Taking 0.1% as hydrogen, the estimated mass of hydrogen in the ice layer is 0.025 $M_\oplus$. Thus for the RIG the ratio of hydrogen in the atmosphere to that in ice is 305 giving the value of D/H shown in Table 15.6.

The characteristics of the RIG are clearly those that are associated with an ice giant. The large tilt is brought about by the direct transfer of angular momentum by the impact plus the huge tidal effect Bellona exerts on the PIG, especially during the close approach of the two bodies. By varying the parameters of the model other outcomes are possible, with more or less mass and radius and with a greater or lesser tilt of the spin axis.

This scenario for explaining the characteristics of the ice giants is speculative and appropriate to a CT origin of the planets. Alternative scenarios may be appropriate to other planet-forming theories and may be equally successful in explaining the characteristics of the ice giants.

15.23. Isotopic Anomalies in Meteorites

The presence of a hot silicate vapour in the early Solar System, the conclusion from studies of meteorites that stimulated the rebirth of nebula ideas, would be a natural consequence of a planetary collision. Another interesting characteristic of some meteorites is that the isotopic compositions for many elements are very different from those of terrestrial material; the differences are denoted as *isotopic anomalies.* These are described here for carbon, nitrogen, oxygen and neon, but there are many others.

The ratio of the two stable isotopes of carbon, ${}^{12}_{6}\mathrm{C}$:${}^{13}_{6}\mathrm{C}$, in terrestrial material is 89.9:1. In mineral grains of silicon carbide, SiC, in some chondritic meteorites the ratio is much less, down to 20:1, giving what is known as *heavy carbon.* The usual explanation given by meteoriticists is that grains containing heavy carbon drifted into the Solar System from six or more distant carbon stars — red giants with excess carbon in their atmospheres — each with carbon of different heaviness.

Silicon carbide also contains nitrogen trapped in interstices between grains. The ratio of the two stable nitrogen isotopes on Earth, ${}^{14}_{7}\mathrm{N}:{}^{15}_{7}\mathrm{N}$, is 270:1. Most SiC-derived nitrogen is *light nitrogen* with ratios up to 2,000:1 but, rarely, *heavy nitrogen* occurs with ratios down to 50:1.

The three stable isotopes of oxygen have terrestrial ratios

$$ {}^{16}_{8}\mathrm{O} : {}^{17}_{8}\mathrm{O} : {}^{18}_{8}\mathrm{O} = 0.9527{:}0.0071{:}0.0401. $$

This mixture, known as SMOW (Standard Mean Ocean Water), also occurs in Moon rocks. Two kinds of meteorite contain oxygen isotopic ratios that cannot be explained by processing terrestrial oxygen. These are ordinary chondrites and carbonaceous chondrites, the latter being stony meteorites that contain volatile material. The oxygen anomalies can be explained as the addition of different amounts of pure, or nearly pure, ${}^{16}_{8}\mathrm{O}$ to SMOW. The usual explanation again involves grains drifting across interstellar space and entering the Solar System carrying excess ${}^{16}_{8}\mathrm{O}$.

The final example is neon. The three stable isotopes have terrestrial ratios

$$ {}^{20}_{10}\mathrm{Ne}:{}^{21}_{10}\mathrm{Ne}:{}^{22}_{10}\mathrm{Ne} = 0.905{:}0.003{:}0.092. $$

Neon atoms are trapped in atomic-size cavities, but can be released by heating the meteorite. If neon is found in meteorites then they could not have been substantially heated after the neon was incorporated. The same is true for other gases trapped in the interstices of meteorite grains.

Isotopic compositions of neon in meteorites are very variable and it has been deduced that they come from admixtures of three neon sources with different compositions. However, some meteorites contain pure, or almost pure, ${}^{22}_{10}\mathrm{Ne}$ that is a 9.2% component of terrestrial neon. This anomalous neon is called *neon-E.* Since

$^{22}_{10}$Ne cannot be separated from a mixture of isotopes some other explanation is required. The only stable isotope of sodium is $^{23}_{11}$Na, but there is a radioactive isotope, $^{22}_{11}$Na, which decays into $^{22}_{10}$Ne. One suggested scenario is that just before solar-system formation a nearby supernova produced $^{22}_{11}$Na that was incorporated with stable sodium, in minerals. The $^{22}_{11}$Na decayed and the resultant $^{22}_{10}$Ne was trapped within the mineral grains. A problem with that scenario is that $^{22}_{11}$Na has a half-life of 2.6 years so that, after production in a supernova, it has to be incorporated into a *cool* solid body within a period of less than 30 years, otherwise not much of the radioactive sodium will remain. This puts a very tight constraint on the timing of the supernova.

Holden and Woolfson (1995) examined the effect of subjecting a mixture of iron, silicates and ices, with a *D*/*H* ratio that of Venus (0.016), to a triggering temperature of 3×10^6 K. The rates given by Fowler, Caughlan and Zimmerman (1967, 1975) for 548 nuclear reactions were used and 40 decay processes were incorporated in their calculation. All possible cooling factors were included; iron, which took no part in reactions, was a coolant and ionization of material, implemented by a solution of the Saha equations (Zel'dovich and Raiser, 1966), greatly increased the number of particles present to share the generated energy. The outcome was a nuclear explosion, the products of which explained a number of important light-atom isotopic anomalies, including those described above. The final temperature was well in excess of 5×10^8 K.

Another calculation by Woolfson (2011c), using a lesser *D*/*H* ratio of 0.01, gave similar results, explaining all anomalies for carbon, nitrogen, oxygen, neon, magnesium, aluminium and silicon. This single explanation replaced a number of *ad hoc* explanations for individual anomalies.

A large quantity of $^{13}_{6}$C, and radioactive $^{13}_{7}$N that decays to $^{13}_{6}$C with a half-life of 9.97 min, was produced, which explained the full range of heavy-carbon observations. Figure 15.26 shows the concentrations of $^{12}_{6}$C and $^{13}_{6}$C (including $^{13}_{7}$N contribution). When the temperature exceeds 3×10^8 K there is a sharp increase in the amount of $^{13}_{6}$C.

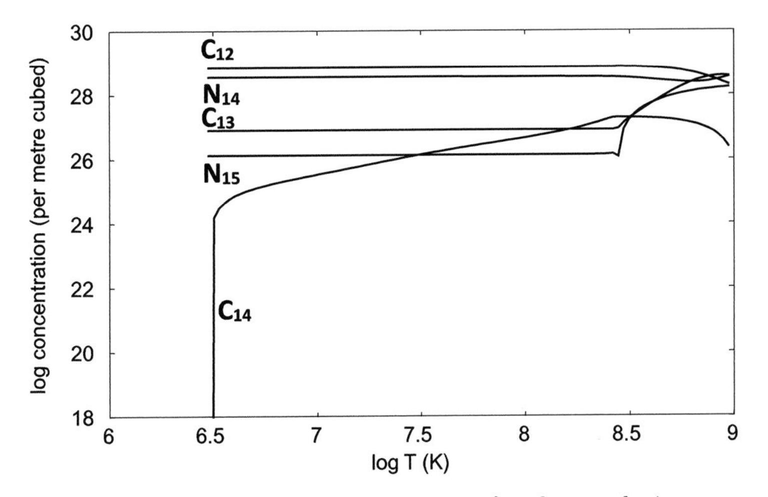

Figure 15.26. The concentrations of isotopes of carbon and nitrogen as the temperature within the nuclear-reaction region increases.

There is a small reduction in the amount of $^{14}_{7}$N, including the contribution of $^{14}_{8}$O that quickly decays to $^{14}_{7}$N, as the explosion progresses, although it picks up at very high temperatures (Figure 15.26). In the longer term the concentration of $^{14}_{7}$N is augmented by $^{14}_{6}$C decay with a half-life of 5,739 years. Starting with $^{14}_{7}$N and $^{15}_{7}$N trapped inside the grains once they were cool enough to retain the gas, $^{14}_{6}$C, which had been part of the silicon-carbide host, decayed, boosting the amount of $^{14}_{7}$N, so giving light nitrogen. If the original amount of the two stable nitrogen isotopes was not very great then the $^{14}_{6}$C contribution can make a large proportional change to the $^{14}_{7}$N concentration.

Towards the end of the explosion there is an almost 100-fold increase in the amount of $^{15}_{7}$N present, produced by reactions involving heavier elements. These elements may not be uniformly distributed so that $^{15}_{7}$N-rich pockets can form. Even after the decay of $^{14}_{6}$C, the result can be the occasional occurrence of heavy nitrogen.

Figure 15.27 shows the variation of stable oxygen isotopes during the explosion. The $^{17}_{8}$O and $^{18}_{8}$O concentrations include contributions from the fluorine radioactive isotopes, $^{17}_{9}$F and $^{18}_{9}$F, which quickly decay to $^{17}_{8}$O and $^{18}_{8}$O, respectively. Above about 5×10^{8} K, the

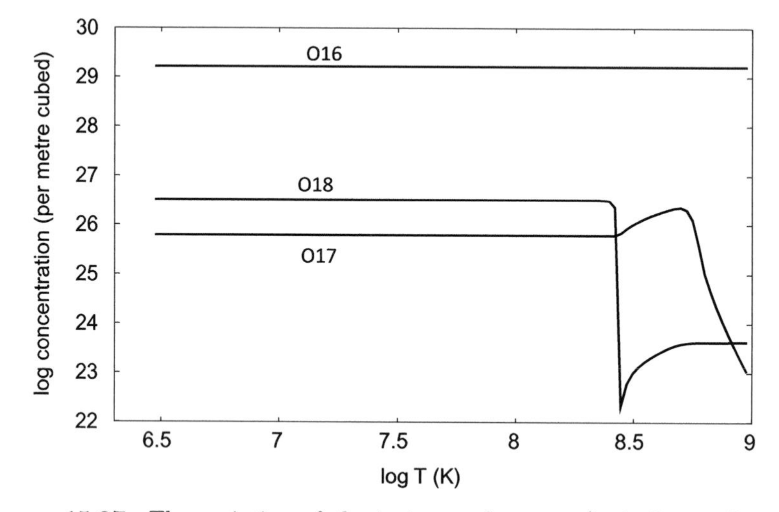

Figure 15.27. The variation of the isotopes of oxygen (including radioactive fluorine) with temperature.

concentrations of $^{17}_{8}$O and $^{18}_{8}$O greatly diminish leaving virtually pure $^{16}_{8}$O that, mixed with SMOW in various proportions, gives the oxygen isotopic anomaly.

A sufficient quantity of $^{22}_{11}$Na was produced in the explosion to explain the production of neon-E. The scale of a planetary collision is small by astronomical standards and the formation of cool grains, which would retain $^{22}_{10}$Ne, would take place within hours or days (Woolfson, 2011d).

15.24. An Overview of the Capture Theory

The origin of the CT was at a time before it was known that the DES of a cluster occurred. It was also at a time that the existence of exoplanets was unknown so the CT had then the status of being a rare occurrence that explained a rare entity, the Solar System. However, the knowledge that exoplanets are common, closely followed the discovery of the DES so the CT evolved into a model of a common mechanism — the interactions within a DES — to explain commonly-occurring entities, planetary systems.

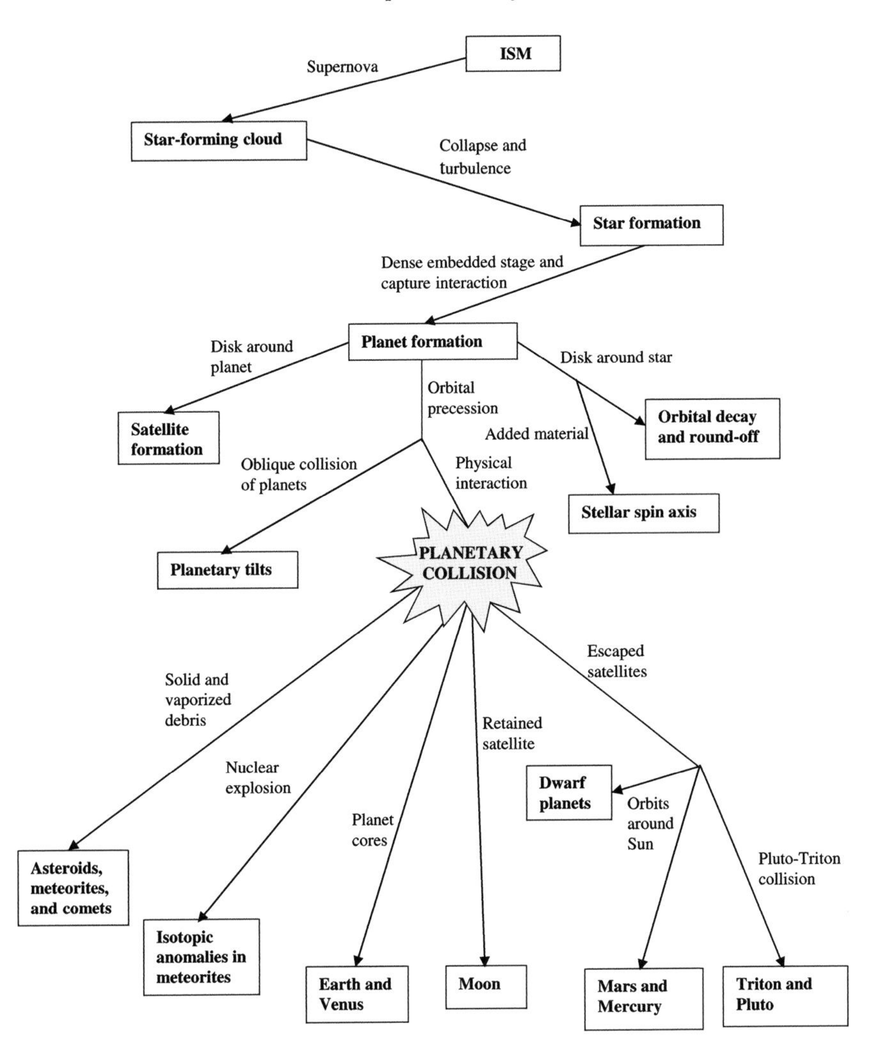

Figure 15.28. Consequences of the Capture theory and a planetary collision.

Both the NT and the CT have circumstellar disks playing important roles, in the case of the NT both as sources of planetary material and as a medium for migration but for the CT just the latter role. Again, the disks have different origins; for the NT they are derived from the process that produced stars and for the CT derived from the process that produced planets.

Type II inward migration is faster for more massive planets, which is consistent with Bellona and Enyo reaching the terrestrial region while other planets were still further out. This increased the likelihood that they would collide, an event that explains many features of the Solar System in terms of both what the system contained before the collision and the products of the collision. The discovery of enhanced D/H ratios in grains and the consequent nuclear explosion added greatly to the energy available to disperse the products of the collision. This nuclear explosion could only have taken place with concentration of deuterium and it gives a common explanation for many isotopic anomalies, replacing a number of *ad hoc* explanations.

Figure 15.28 shows as a flow diagram the various causally-related processes that explain the formation of planets and features of the Solar System.

Problem 15

15.1 A circumplanetary disk has a mass of 1.5×10^{27} kg and a radius of 1.5 au. If 1% of it is converted into satellitesimals then what is their mean areal density?

15.2 A satellite of mass 10^{23} kg and mean density $3 \times 10^3\,\mathrm{kg\,m^{-3}}$ is produced at distance 4×10^5 km from the planet of mass 2×10^{27} kg. If the areal density of satellitesimals is four times the mean value found in Problem 15.1 then what is the formation time for the satellite? (take $\beta = 4$).

Appendices

Appendix A

Planck's Radiation Law and Quantum Physics

The radiation coming from a star is usually regarded as similar to that coming from a *black body*, one that absorbs all the radiation falling on it. An example of a black body is an enclosure with a small aperture in it, as illustrated in Figure A.1. Any radiation entering the enclosure via the aperture will be reflected or reradiated by the walls of the enclosure and be unlikely to leave again. On the other hand some radiation does escape from the enclosure. The aperture is a good approximation to a black body and the radiation escaping from it will be a representative sample of the radiation that exists inside the enclosure. This will depend on the temperature of the enclosure so what we are interested in is the density of radiation at different wavelengths, or frequencies, inside it. This is the problem solved by Max Planck (1858–1947; Figure A.2) in 1900.

Figure A.1. Radiation entering a black body.

Figure A.2. Max Planck.

A.1. The Rayleigh–Jeans Radiation Law

The ultimate aim in this part of the analysis leading to Planck's law of black-body radiation is to find the distribution of frequencies within a cavity. We consider radiation in equilibrium within a cube of side a. For equilibrium we need standing waves inside the cube and the waves must have nodes, i.e. zero amplitudes, at the cavity walls. The three-dimensional wave equation is

$$\frac{\partial^2 f}{\partial x^2} + \frac{\partial^2 f}{\partial y^2} + \frac{\partial^2 f}{\partial z^2} = \frac{1}{c^2}\frac{\partial^2 f}{\partial t^2}. \tag{A.1}$$

Given the boundary conditions the solution is of the form

$$f = C\sin\left(\frac{\pi n_x}{a}x\right)\sin\left(\frac{\pi n_y}{a}u\right)\sin\left(\frac{\pi n_z}{a}z\right)\sin\left(\frac{2\pi ct}{\lambda}\right), \tag{A.2}$$

where the n's are positive integers and C an arbitrary constant. Inserting this solution into (A.1) gives the condition

$$n_z^2 + n_y^2 + n_z^2 = \left(\frac{2a}{\lambda}\right)^2 = \left(\frac{2\nu a}{c}\right)^2. \tag{A.3}$$

For a frequency ν to be present in the cavity there have to be values of n_x, n_y and n_z satisfying equation (A.3).

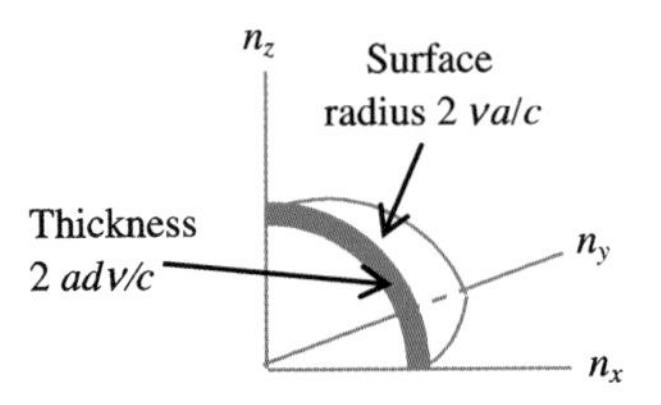

Figure A.3. The volume giving the number of modes between ν and $\nu + d\nu$.

We are now going to consider the number of solutions with frequencies between ν and $\nu + d\nu$. In Figure A.3 n_x, n_y and n_z are plotted on orthogonal axes and the number of radiation modes with frequencies between ν and $\nu + d\nu$ is twice the volume in the quadrant corresponding to positive n's between radii $2\,\nu a/c$ and $2(\nu + d\nu)a/c$. The extra factor of two comes from the fact that light consists of transverse waves and that there are two possible independent directions of polarization of the waves. This gives

$$N(\nu)d\nu = 2 \times \frac{4\pi}{8}\left(\frac{2a}{c}\right)^3 \nu^2 d\nu = \frac{8\pi a^3 \nu^2}{c^3} d\nu. \tag{A.4}$$

We now divide (A.4) by a^3 to give the number of independent modes per unit volume per unit frequency and take each mode of vibration as contributing energy kT — a result coming from statistical mechanics. This gives the Rayleigh–Jeans radiation law for the radiation density per unit frequency interval within an enclosure at temperature T.

$$E(\nu) = \frac{8\pi kT\nu^2}{c^3}. \tag{A.5}$$

The law is clearly flawed since it indicates that the energy density increases with frequency without limit. This problem with the law is termed the *ultraviolet catastrophe.*

A.2. The Planck Radiation Law

The failed Rayleigh–Jeans radiation law had assumed that there was energy kT associated with each independent wave mode within the cavity. For a wave motion the energy is proportional to the square of the amplitude and Planck introduced the idea that a wave mode could have associated with it more than a single energy but that the allowed energy was quantized of the form

$$\varepsilon_\nu = nh\nu, \tag{A.6}$$

where n is a non-negative integer and h a constant now known as Planck's constant. We have to remember that the time that Planck introduced his concept of the quantization of energy was a time of classical physics. Planck introduced his quantization proposal without any justification other than that it worked — he used the term *resonators* for the entities that would have quantized energy. In 1905, in explaining the photoelectric effect, Einstein showed that, in some circumstances, light could behave like a stream of particles — photons — that had energy $h\nu$ for light of frequency ν, We know from the Pauli exclusion principle that not more than one fermion can occupy any state of a quantum-mechanical system — it can be vacant (as for an ion) or have single occupancy. However, photons are *bosons*, with integer spin ($\pm\hbar$) and a state can be occupied by any number of bosons. Thus for each independent mode of vibration, that we can think of as a possible state of the system, there can be any number of equivalent photons, including zero. Translating this into electromagnetic-wave terms the energy of the vibrational mode, which is proportional to the square of the amplitude, can be $nh\nu$, for any positive n and also zero, the last possibility meaning that this mode is absent.

We now have to consider the probability for any particular energy of a mode with frequency ν.

Since they are bosons, photons should obey Bose–Einstein statistics, for which the probability that a state has energy E is proportional to

$$p_{BE}(E) = \frac{1}{\exp\left(\frac{E}{kT}\right) - 1}. \tag{A.7}$$

This has the characteristic that the actual probability

$$P_{BE}(E_n) = \frac{p(E_n)}{\sum_{n=0}^{\infty} p(E_n)}, \tag{A.8}$$

becomes indeterminate for $E = 0$, the energy corresponding to a vacant mode of vibration.

The alternative is to assume that considered as an electromagnetic vibration the weighting factor to be used is the Boltzmann factor, which comes from statistical mechanics, which is proportional to

$$p_B(E) = \exp\left(-\frac{E}{kT}\right). \tag{A.9}$$

Using this, the average energy for a mode of frequency ν is

$$\langle \varepsilon_\nu \rangle = \frac{\sum_{n=0}^{\infty} nh\nu \exp\left(-\frac{nh\nu}{kT}\right)}{\sum_{n=0}^{\infty} \exp\left(-\frac{nh\nu}{kT}\right)}. \tag{A.10}$$

To simplify the analysis we write
$y = \exp\left(-\frac{h\nu}{kT}\right)$ that gives

$$\langle \varepsilon_\nu \rangle = h\nu y \frac{\sum_{n=1}^{\infty} ny^{n-1}}{\sum_0^{\infty} y^m} = h\nu y \frac{P}{Q}. \tag{A.11}$$

The lower summation, Q, is a geometrical progression of the form $a + ar + ar^2 + ar^3 + \ldots$, with leading term 1 and the multiplier for successive terms y. When $r < 1$ the infinite sum is

$$Q = \frac{a}{1-r} = \frac{1}{1-y}. \tag{A.12}$$

We also note that

$$P = \frac{dQ}{dy} = \frac{1}{(1-y)^2}. \tag{A.13}$$

Combining (A.10), (A.11) and (A.12) we find

$$\langle \varepsilon_\nu \rangle = \frac{h\nu y}{1-y} = h\nu \frac{\exp\left(-\frac{h\nu}{kT}\right)}{1 - \exp\left(-\frac{h\nu}{kt}\right)} = \frac{h\nu}{\exp\left(\frac{h\nu}{kt}\right) - 1}. \tag{A.14}$$

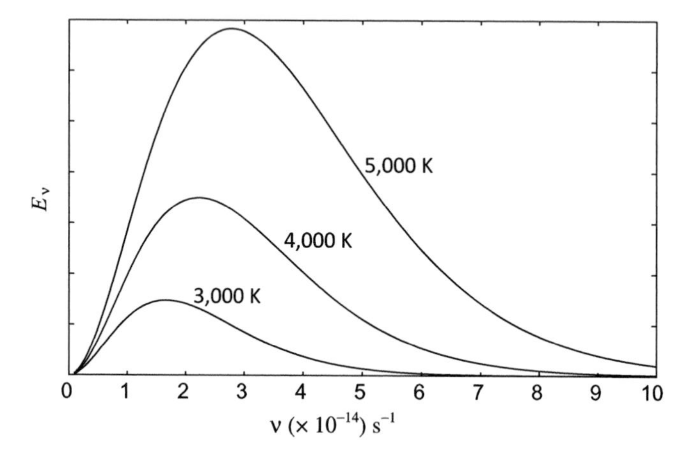

Figure A.4. Planck radiation curves.

Replacing kT by $\langle \varepsilon_\nu \rangle$ in (A.5) gives the Planck equation for black-body radiation

$$\langle E_\nu \rangle = \frac{8\pi h\nu^3}{c^3} \frac{1}{\exp\left(\frac{h\nu}{kT}\right) - 1}. \tag{A.15}$$

The form of this function is shown for three temperatures in Figures 2.9 and A.4.

There are other approaches to the derivation of Planck's law, one of which assumes that the quantization of energy come from harmonic oscillators with energies $(n + 1/2)h\nu$. However, every derivation has some associated conceptual problem and the derivation given here is probably the simplest.

Problem A

A.1 Transform A(14) into a distribution in terms of wavelength λ. i.e. such that $E(\lambda)d\lambda$ is the proportion of the energy with wavelengths between λ and $\lambda + d\lambda$. Find the wavelength of the peak of this distribution, λ_{max}, and hence show that $\lambda_{\text{max}} T$ is a constant (Wien's displacement law).

Appendix B

The Relativistic Doppler Effect

It was generally believed in the 19th century that all forms of wave motion, such as sound or electromagnetic radiation, had to travel in a medium of some kind. This was certainly true for sound; an electric bell ringing in an evacuated bell-jar cannot be heard. It was proposed that the medium for electromagnetic radiation, in particular light, was the *ether*, which was extremely tenuous but very elastic, which properties explained the high speed of light. Just as a boat, seen from the shore, is slower when moving against the flow of water than when moving with it, so it was expected that the observed speed of light would depend on its motion relative to the ether. However, an experiment carried out in 1887 by the American scientists, Albert Michelson and Edward Morley, failed to detect any variation in the speed of light regardless of its direction of motion.

In 1905, Albert Einstein published his Theory of Relativity that was founded on two postulates.

(i) All inertial frames of reference (i.e. frames of reference not experiencing forces) are equally valid for describing events in the Universe and the laws of physics are the same as seen from any one of them.

(ii) There is no experiment that can distinguish one inertial frame from another and, in particular, the speed of light is the same in all inertial frames.

This is the basis of Special Relativity Theory and of the Relativistic Doppler Effect.

B.1. A Non-Relativistic Moving Clock

The word 'clock' conjures up the vision of an analogue or digital device for recording the passage of time. However, any entity that changes in a predictable way with time, or has a periodic variation at a constant rate can be considered as equivalent to a clock. This would include a light wave, the peaks of which vary in a periodic way, and the separation in time of two peaks can be considered as equivalent to the interval between two ticks of a mechanical clock. Hence, any property of conventional clocks that we find can be reinterpreted in terms of the interval between successive peaks of a light wave — the inverse of its frequency.

Figure B.1 shows an observer O with a clock, recording time from an identical synchronized clock moving with speed Vaway from O. When the moving clock is at position A it records time 0 and when O sees time 0 on the moving clock the time on his clock is D/c where c is the speed of light. After a further tine τ the moving clock has moved to position B, a distance $V\tau$ from A and the passage of time recorded on O's clock when he sees time τ on the moving clock is $\tau + (D + V\tau)/c - D/c$. Thus when O sees the passage of time τ on the moving clock he sees the passage of time $\tau(1 + V/c)$ on his own clock. In general the ratio of the time seen by O on the moving clock, t_m to that seen on his own clock, t_O, is given by

$$\frac{t_m}{t_O} = \frac{1}{1 + V/c}. \tag{B.1}$$

If the moving clock is receding then V is positive and the moving clock appears to run slowly. Conversely if the moving clock is

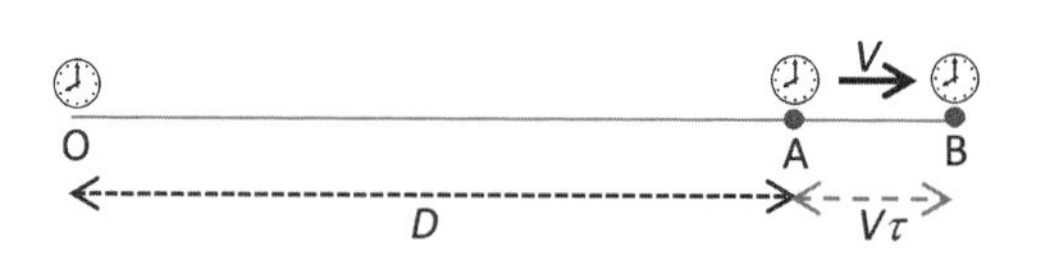

Figure B.1. An observer and a moving clock.

approaching O then V is negative and the moving clock appears to be running fast.

Actually, what we have done is to show the result for the classical Doppler Effect. For a wave motion, time can be recorded by fluctuations of the wave and a slowly running clock is equivalent to having a lower frequency — and hence a longer wavelength. Converting to wavelength

$$\frac{\lambda_{\mathrm{m}}}{\lambda_O} = 1 + \frac{V}{c}. \tag{B.2}$$

This means that when the source moves away from O the observed wavelength is increased and is shortened when the source is approaching the observer. In terms of the change of wavelength $d\lambda = \lambda_m - \lambda$

$$\frac{d\lambda + \lambda_{\mathrm{o}}}{\lambda_O} = 1 + \frac{V}{c},$$

or

$$\frac{d\lambda}{\lambda_O} = \frac{V}{c},$$

which is the classical Doppler Effect equation.

By virtue of its motion, the moving clock is seen to be running either quickly or slowly. However, by applying the laws of classical physics the observer would be able to deduce that the moving clock was recording time at the same rate as his own clock.

B.2. A Relativistic Moving Clock

Now another experiment, illustrated in Figure B.2, will be described in which the postulate that all inertial observers see the same speed of light will be invoked.

In this scenario the observer O sees a brief light pulse follow the path OAB, being reflected by the mirror at A. At the time the light pulse leaves O a clock starts moving along the line OB at a speed V such that the clock reaches B at the same time as the light pulse. The observer sees the light pulse travel a distance $2L$ ($L = \mathrm{OA} = \mathrm{AB}$) at a speed c and so in his frame of reference the time taken, and seen

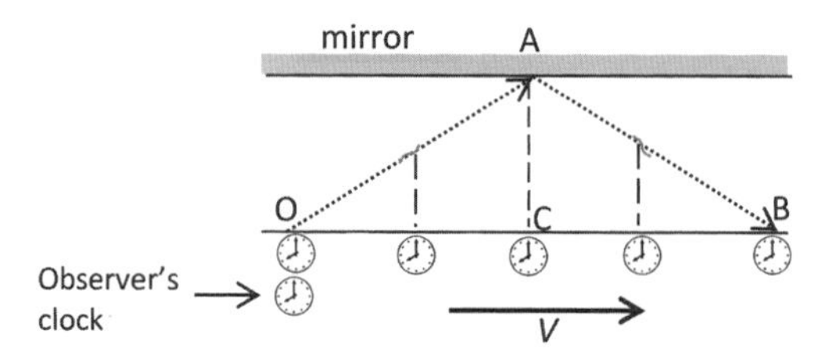

Figure B.2. An experiment with a moving clock using relativity theory.

on his clock, is

$$t_O = \frac{2L}{c}, \tag{B.3}$$

where L is the distance OA = AB. Since the observer sees the moving clock reaching B at the same time as the light pulse then

$$\mathrm{OC} = \frac{LV}{c}. \tag{B.4}$$

Now we consider what is seen by an observer travelling with the moving clock. He always sees the light pulse moving perpendicular to the mirror so that in his frame of reference it travels a distance

$$2\mathrm{AC} = 2\sqrt{\mathrm{OA}^2 - \mathrm{OC}^2} = 2L\sqrt{1 - \frac{V^2}{c^2}}.$$

Since the speed of light is c, the time for the pulse of light to reach B, and seen by the second observer on his clock is

$$t_m = \frac{2L}{c}\sqrt{1 - \frac{V^2}{c^2}},$$

or

$$\frac{t_m}{t_O} = \sqrt{1 - \frac{V^2}{c^2}}. \tag{B.5}$$

By a similar transformation that gave equation (B.2) from (B.1)

$$\frac{\lambda_m}{\lambda_O} = \frac{1}{\sqrt{1 - V^2/c^2}}. \tag{B.6}$$

Now the time seen by O on the moving clock when the light pulse reaches B must be the same as is seen by the second moving observer; consider that the clock face can only be seen when illuminated by the light pulse. The implication of equation (B.5) is that a moving clock runs slowly, regardless of whether it is moving towards or away from the observer. It is not that the observer sees the moving clock running slowly but, by the application of some classical physics could deduce that the two clocks were actually recording time at the same rate. In O's frame of reference the moving clock is running at a slower rate than his own clock.

B.3. The Relativistic Doppler Effect Equation

The observation of a moving clock involves two factors, the first being just due to the change of distance between two readings of time, as explained in Section B.1, and the second due to a slowing of the rate at which the moving clock is running in the inertial frame of the observer, as explained in Section B.2. These factors, given by equations (B.1) and (B.6) so that the ratio of the time seen by O on his own clock to that seen on the moving clock is

$$\frac{t_{\rm m}}{t_{\rm O}} = \frac{1}{1+V/c} \times \sqrt{1 - \frac{V^2}{c^2}} = \sqrt{\frac{1-V/c}{1+V/c}}. \tag{B.7}$$

We now write $\beta = V/c$, i.e. speed as a fraction of the speed of light, and express (B.7) in terms of the wavelength of an electromagnetic wave to give $\frac{\lambda_m}{\lambda_O} = \sqrt{\frac{1+\beta}{1-\beta}}$ and then, with $d\lambda = \lambda_m - \lambda_{\rm O}$, we define the resulting *redshift* as

$$z = \frac{d\lambda}{\lambda_{\rm O}} = \sqrt{\frac{1+\beta}{1-\beta}} - 1. \tag{B.8}$$

The quantity z, the redshift, is the change of wavelength, positive or negative, in units of the original wavelength and is shown as a function of β in Figure B.3. It will be seen that for small magnitudes of β the classical and relativistic results are similar.

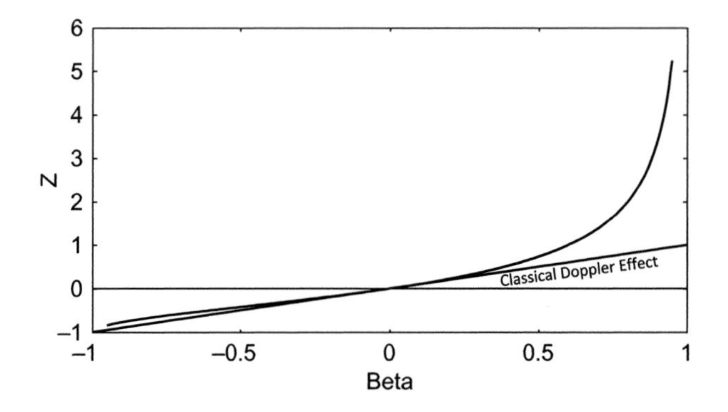

Figure B.3. The Z–β relationship for the relativistic Doppler Effect.

Problem B

B.1 A spectral line is seen with a redshift of 0.4. What is the speed of the radiating body?

Appendix C

Energy Production in Stars

For a body of mass greater than about 0.013 $M_\odot$, in the Kelvin–Helmholtz stage of contraction towards the main sequence, an internal temperature of a few million degrees occurs, which is sufficient to trigger D–D reactions. This reaction is $\mathrm{D}(\mathrm{D}, p)\mathrm{T}$ which is equivalent to

$$\mathrm{D} + \mathrm{D} \rightarrow \mathrm{T} + p,$$

where p is a proton and T is tritium, a radioactive isotope of hydrogen containing two neutrons in its nucleus. Once tritium is produced it can take part in other reactions such as $\mathrm{T}(\mathrm{D}, n)^4\mathrm{He}$ and $\mathrm{T}(\mathrm{T}, 2n)^4\mathrm{He}$. However, the concentration of deuterium in hydrogen is small, with $\mathrm{D/H} = 2 \times 10^{-5}$, and the temperature rise due to deuterium-based reactions is too small to trigger further reactions. If only deuterium reactions occur then the body is defined as a *brown dwarf* and never becomes a main-sequence star.

If the mass of the body is greater than 0.075 $M_\odot$ then the internal temperature of the body reaches the level at which reactions involving hydrogen can take place. Once this happens the temperature rises to the point where the energy generated equals the energy lost by radiation from the body, which then becomes a star on the main sequence. With an abundance of hydrogen fuel available, this is a long-lasting state.

C.1. Proton–Proton Reactions from a Classical Viewpoint

The central temperature of the Sun is about 15 million Kelvin, at which temperature the reaction $\mathrm{H}(p, e^{+} + \nu)\mathrm{D}$ takes place, where e^{+} is a positron and ν is a neutrino. It is customary to write H outside the bracket and p within it although they actually represent the same kind of particle, a hydrogen nucleus or a proton. We now consider the general conditions under which a nuclear reaction can occur between two protons.

The proton radius is of order 10^{-15} m, outside of which distance it gives a Coulomb potential that would repel another proton. However, if two protons were to approach more closely than this then another strong force of attraction occurs. Figure C.1 shows the potential energy as a function of the distance between two protons. There is a potential barrier, of height E_p, that must be overcome by an approaching proton; once this barrier is crossed then the strong force of attraction takes over and a nuclear reaction may then occur. The maximum height of the barrier is given by

$$E_{\mathrm{p}} = \frac{e^2}{4\pi\varepsilon_0 r_{\mathrm{p}}}, \tag{C.1}$$

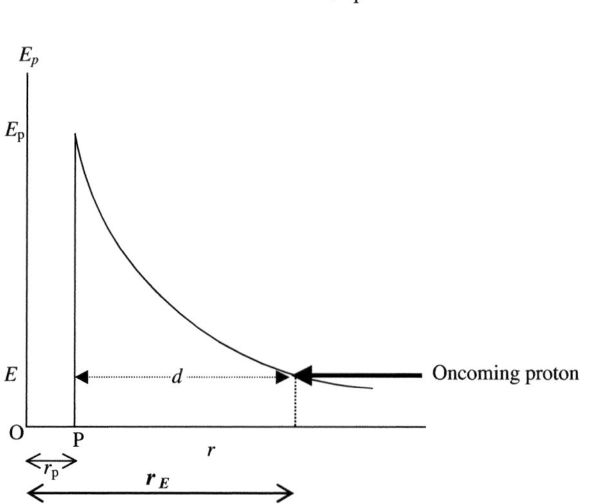

Figure C.1. Representation of a proton of energy $\boldsymbol{E}$ approaching a Coulomb barrier of maximum height $\boldsymbol{E_p}$.

where e is the proton charge (1.602×10^{-19} C) and ε_0 is the permittivity of free space (8.854×10^{-12} F m^{-1}). With $r_p = 10^{-15}$ m this potential barrier is 2.31×10^{-13} J or 1.44×10^6 eV. If the translational kinetic energy of the approaching particle is less than this then, according to classical ideas, it will undergo an elastic repulsion. We may translate this energy into a temperature by

$$E_p - 3kT/2,$$

and the corresponding temperature is 1.1×10^{10} K, which is three orders of magnitude greater than the Sun's central temperature. According to classical theory proton–proton reactions cannot take place within the Sun — but they do.

C.2. An Approximate Quantum-Mechanical Approach

The explanation for proton–proton reactions comes from the quantum-mechanical mechanism of *tunnelling*. Figure C.2 shows a particle of mass m and kinetic energy E meeting a rectangular barrier of height E_m ($>E$). From the application of quantum mechanics it is found that the particle has a probability of penetrating the barrier given by

$$P_{\text{pen}} = \exp\left\{-2\sqrt{\frac{2m\,(E_m - E)}{\hbar^2}}\,d\right\}, \tag{C.2}$$

where $\hbar = h/2\pi$ and h is *Planck's constant*, 6.626×10^{-34} J s.

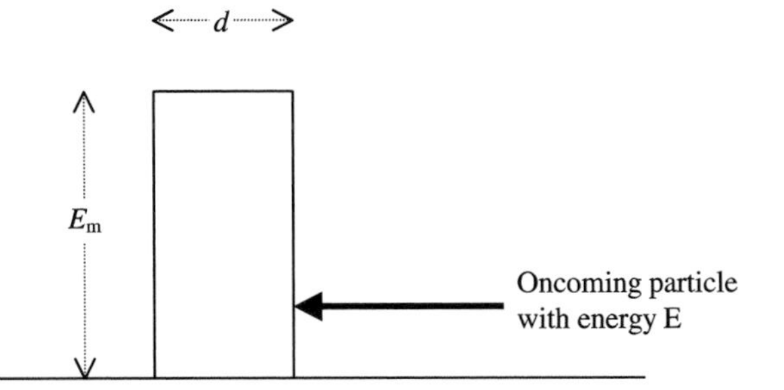

Figure C.2. A particle of energy E approaching a rectangular potential barrier of height E_m and thickness d.

Having penetrated the barrier, the particle emerges from it with the same energy, E, with which it entered the barrier. All other particles are reflected backwards with energy E, as in the classical case.

For a pair of protons, as an approximation we substitute $E_m = \frac{1}{2}E_p$ in (C.2), and we use for m the *reduced mass* of the two-particle system $m_r = \frac{m_1 m_2}{m_1+m_2}$ ($=\frac{1}{2}\, m_p$, where m_p is the proton mass, in this case).

We also take d as the distance shown in Figure C.1, which together will give a rather rough estimate of the probability of barrier penetration. From Figure C.1

$$d = \frac{e^2}{4\pi\varepsilon_0 E} - r_p, \tag{C.3}$$

where r_p is 10^{-15} m, the radius of the proton.

For the central temperature of the Sun, 1.5×10^7 K, the mean translational energy of a proton is $3kT/2$ or 3.1×10^{-16} J. There will be many protons with three times the mean energy, or more, so we take $E = 10^{-15}$ J. From (C.3) we find $d = 2.29 \times 10^{-13}$ m. The value of E_m, found from (C.1) with $r = r_p$ is 2.31×10^{-13} J. With these values (C.2) gives $P_{\text{pen}} \approx 10^{-37}$. The model we have used is very crude but the value found for P_{pen} indicates that the fusion of hydrogen takes place very slowly in the interior of the Sun and other stars. To get a quantitative assessment of proton–proton reaction rates we now develop a more detailed theory.

C.3. A More Precise Quantum-Mechanical Approach

In the approximate approach to tunnelling we took averages of various quantities. For a more precise analysis we need to take into account the distributions of quantities rather than just their average values.

C.3.1. *The Distribution of the Relative Energies of Protons*

Since $E \ll E_m$ the value of $(E_m - E)$ is little effected by large relative changes of E. However, $r_p \ll d$ so, from Figure C.1, it can be seen that the value of d is inversely proportional to E. Thus, from equation (C.2) it is evident that penetration increases rapidly with particle energy. Since the number of high-energy particles increases rapidly with increasing energy so, clearly, the energy distribution is important.

What is relevant in two-particle interactions is the relative energy of pairs of particles, i.e. the kinetic energy of one particle as seen by the other, considered at rest. This distribution is the Maxwell–Boltzmann distribution that gives the proportion of particles with relative energy between E and $E + dE$ as

$$P(E)dE = 2\pi \left(\frac{1}{\pi kT}\right)^{3/2} E^{1/2} \exp\left(-\frac{E}{kT}\right) dE. \tag{C.4}$$

C.3.2. *The Rate of Making Close Approaches*

If we were considering the rate of collisions of a large number of classical particles — for example, a large number of billiard balls moving randomly on a billiard table — the factors governing the collision rate for a particular ball would be the speeds of the balls and their radii. However, the behaviour of small bodies like protons is governed by the rules of quantum mechanics and we must take account of the fact that in some circumstances they have wave-like properties. The wavelength of a quantum-mechanical particle, λ, with momentum $p(=mv)$ is given by the *de Broglie hypothesis* as

$$p = \frac{h}{\lambda} \quad \text{or} \quad \lambda = \frac{h}{p} = \frac{h}{mv}.$$

We take the close-approach distance equal to the de Broglie wavelength for a particle of mass m_r moving at the relative speed of the two particles, v_r, i.e.

$$\lambda = \frac{h}{m_r v_r} = \frac{h}{\sqrt{2Em_r}}. \tag{C.5}$$

We are going consider a situation in which we have N protons within a volume V. We then calculate the probability per unit time that a particular pair will approach to within a distance λ, the quantum-mechanical equivalent of the classical distance r_p, and then the probability that this will lead to tunnelling and a nuclear reaction. Finally, by considering all possible pairs, we shall calculate the number of reactions per unit time per unit mass of hydrogen.

Relative to one of a pair of protons the other sweeps out a volume $\pi\ \lambda^2 v_r$ per unit time such that if the other falls within this volume then a close approach will have taken place. The probability of this is

$$P_{ap} = \frac{\pi\lambda^2 v_r}{V} = \pi\frac{h^2}{V}\sqrt{\frac{1}{2Em_r^3}}, \tag{C.6}$$

where we have used $E = {}^1\!/\!{}_2 m v_r^2$ or $v_r = \sqrt{\frac{2E}{m_r}}$.

The close approach distance λ defined here is much larger than r_p. Now we find the probability that the approach leads to a tunnelling event.

C.3.3. *The Tunnelling Probability*

Consider the potential barrier shown in Figure C.1 as divided up into s narrow strips, each of width Δx. The probability that the particle penetrates the ith strip is, from equation (C.2),

$$P_i = \exp\left\{-2\sqrt{\frac{2m_r\left(E_i - E\right)}{\hbar^2}}\Delta x\right\}.$$

The probability that the particle penetrates the whole barrier is the product of the probabilities for the s strips

$$P_{\text{pen}} = \prod_{i=1}^{s} P_i = \prod_{i=1}^{s} \exp\left\{-2\sqrt{\frac{2m_r\left(E_i - E\right)}{\hbar^2}}\,\Delta x\right\}. \tag{C.7}$$

Taking the natural logarithm of each side of (C.7)

$$\ln(P_{\text{pen}}) = -2\sqrt{\frac{2m_r}{\hbar^2}}\sum_{i=1}^{s}\sqrt{(E_i - E)}\Delta x. \tag{C.8}$$

Taken to the limit, with an infinite number of strips

$$\ln(P_{\text{pen}}) = -2\sqrt{\frac{2m_r}{\hbar^2}} \int_{r_p}^{r_E} \sqrt{(E_x - E)}dx, \tag{C.9}$$

where r_E and r_p are shown in Figure C.1.

Since $E_x = \frac{e^2}{4\pi\varepsilon_0 x}$ we have $\frac{dx}{dE_x} = -\frac{e^2}{4\pi\varepsilon_0 E_x^2}$ and changing the variable to E_x gives

$$\ln(P_{\text{pen}}) = \frac{e^2}{2\pi\varepsilon_0}\sqrt{\frac{2m_r}{\hbar^2}} \int_{E_p}^{E} \frac{\sqrt{E_x - E}}{E_x^2} dE_x.$$

Substituting $E_x = E\sec^2\theta$ in the integral with the condition $E_p \gg E$ gives the value of the integral as $-\pi/(2E^{1/2})$ so that

$$\ln(P_{\text{pen}}) = -\sqrt{\frac{e^4 m_r}{8\varepsilon_0^2\hbar^2}} \frac{1}{E^{\frac{1}{2}}}. \tag{C.10}$$

We now define the *Gamow energy*

$$E_G = \frac{e^4 m_r}{8\varepsilon_0^2\hbar^2},$$

which is about 493 keV. In terms of the Gamov energy (C.10) becomes

$$\ln(P_{\text{pen}}) = -\frac{E_G^{1/2}}{E^{1/2}},$$

or

$$P_{\text{pen}} = \exp\left(-\frac{E_G^{1/2}}{E^{1/2}}\right). \tag{C.11}$$

C.3.4. *The Cross-Section Factor*

The cross-section factor for an interaction depends on three separate quantities. The first of these is the size of the target for approach, given by $\pi\lambda^2$, the second is the probability that barrier penetration will take place (Equation (C.11)) and the third is the probability that, after barrier penetration, a reaction will actually occur. This last quantity, the *cross-section factor*, is a slowly-varying function of energy, $S(E)$, that can be determined by laboratory experiments. In practice, to make reasonable measurements the experiments are done at energies much higher than those prevailing in stars and the results extrapolated into the stellar region.

C.3.5. *The Energy Generation Function*

We are now in a position to write down an expression for the probability per unit time that a particular pair of protons will give a reaction. This is

$$P_{\text{pair}} = \int_0^\infty P(E)P_{\text{ap}}P_{\text{pen}}S(E)dE$$

$$= \left(\frac{2\pi}{k^3T^3m_r^3}\right)^{1/2} \frac{h^2}{V} \int_0^\infty S(E)\exp\left(-\frac{E}{kT} - \frac{E_G^{1/2}}{E^{1/2}}\right) dE. \tag{C.12}$$

The total number of interactions for the N protons is $\frac{1}{2}N^2P_{\text{pair}}$, since, the number of pairs of protons is $\frac{1}{2}N(N-1)$ which is approximately $\frac{1}{2}N^2$. The total mass of protons (hydrogen) is Nm_p and the density $\rho = Nm_p/V$. Taking $m_r = \frac{1}{2}m_p$ gives the number of reactions per unit time per unit mass as

$$P_{\text{reac}} = \rho h^2\sqrt{\frac{4\pi}{k^3T^3m_p^7}} \int_0^\infty S(E)\exp\left(-\frac{E}{kT} - \frac{E_G^{1/2}}{E^{1/2}}\right) dE. \tag{C.13}$$

Without detailed knowledge of $S(E)$ the form of the integrand is not defined but over the range of energies of interest $S(E)$ does not vary greatly and so can be taken outside the integration. The exponential term then remains as the integrand and its form is shown in Figure C.3. The combination of terms in the integrand one of which increases with E and the other of which decreases with E leads to

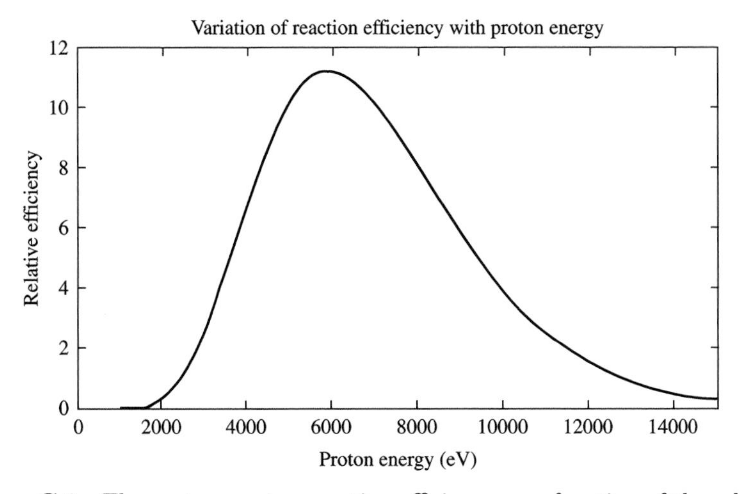

Figure C.3. The proton–proton reaction efficiency as a function of the relative energy of the protons, showing the Gamow peak.

the *Gamow peak*. This occurs when the argument of the exponential is a maximum, when

$$E = \left(1/2 E_G^{1/2} KT\right)^{2/3}. \tag{C.14}$$

For a temperature 1.5×10^7 K, corresponding to the interior of the Sun, E_{max} is about 9.5×10^{-16} J which is about three times the mean energy of the protons. These are the most effective protons. At lower energies there are more protons but they are less effective in producing reactions. At higher energies the protons are more effective in producing reactions but there are fewer of them.

The derivation of the energy generation rate for proton–proton reactions can be generalized to reactions involving other nuclei when the appropriate nuclear charges and masses are incorporated.

C.4. Nuclear Reaction Chains in the Sun

The proton–proton reaction, $\mathrm{H}(p, e^+ + \nu)\mathrm{D}$, is the first step in a chain of reactions leading eventually to the formation of helium-4. The Q *value* of a nuclear reaction is the energy it releases — positive for an exothermic reaction and negative for one that is endothermic.

For this particular reaction it is 1.442 MeV. However, not all the released energy goes into providing heat; the neutrino carries off 0.263 MeV. Neutrinos have a tiny cross-section for interacting with normal matter so that virtually all produced within the Sun leave that body without any interaction.

A series of nuclear reactions in the Sun give the net effect that four protons are converted into a helium nucleus. Since the helium nucleus has less mass than the four protons the mass loss appears as energy according to Einstein's well-known equation $E = mc^2$. The reactions giving this outcome are as follows:

$2 \times (p + p \rightarrow e^+ + \nu + \mathrm{D})$ $\quad Q = 1.442\,\mathrm{MeV}$ with 0.263 MeV taken by neutrino.

$2 \times (D + p \rightarrow \gamma + {}^3\mathrm{He})$ $\quad Q = 5.493\,\mathrm{MeV}$.

${}^3\mathrm{He} + {}^3\mathrm{He} \rightarrow 2p + {}^4\mathrm{He}$ $\quad Q = 12.859\,\mathrm{MeV}$.

These five (2 + 2 + 1) reactions convert four protons into helium with a production of energy

$$2 \times 1.442\,\mathrm{MeV} + 2 \times 5.493\,\mathrm{MeV} \times 12.859\,\mathrm{MeV} = 26.73\,\mathrm{MeV},$$

of which $2 \times 0.263\,\mathrm{MeV} = 0.53\,\mathrm{MeV}$ is carried off by neutrinos.

At the core temperature of the Sun the half-life of a proton is about 8×10^9 years, which explains why the Sun spends so long on the main sequence. By contrast the half-lives for D and ^{3}He are a fraction of a second and 2.4×10^5 years respectively.

There is another process for the conversion of protons into α particles — the CNO cycle. This is described by the following reactions, with the Q value in brackets.

$$\begin{array}{lll}
{}^{12}\mathrm{C} + p & \rightarrow {}^{13}\mathrm{N} + \gamma & (1.944\,\mathrm{MeV}) \\
{}^{13}\mathrm{N} + e^+ & \rightarrow {}^{13}\mathrm{C} + \gamma & (2.221\,\mathrm{MeV}) \\
{}^{13}\mathrm{C} + p & \rightarrow {}^{14}\mathrm{N} + \gamma & (7.550\,\mathrm{MeV}) \\
{}^{14}\mathrm{N} + p & \rightarrow {}^{15}\mathrm{O} + \gamma & (7.293\,\mathrm{MeV}) \\
{}^{15}\mathrm{O} + e^+ & \rightarrow {}^{15}\mathrm{N} + \gamma & (2.761\,\mathrm{MeV} \\
{}^{15}\mathrm{N} + p & \rightarrow {}^{4}\mathrm{He} + {}^{12}\mathrm{C} & (4.965\,\mathrm{MeV})
\end{array}$$

The net result is as before — four protons are converted into a helium nucleus with the production of 26.73 MeV of energy but, for this system, there are no neutrinos to carry off energy. In this process ^{12}C acts as a catalyst; although initially destroyed it is recreated at the end of the chain.

In producing models of stellar structure (Chapter 5) is convenient to have some simple analytical expression. The one normally employed for an interaction between particles A and B is of the form

$$\varepsilon = c\rho X_A X_B T^{\eta}. \tag{C.15}$$

In this expression ε is the energy generation per unit mass of material per unit time, c is a constant, ρ the overall density of the material, X_A is the fractional content of A in the material by mass, T the temperature and η another constant. For the proton–proton reaction $\eta = 4$ but for the CNO reaction $\eta = 17$. The variation of ε with temperature for the two types of reaction is shown in Figure C.4 for solar material. At the temperature of the solar core, $\sim 1.5 \times 10^7$ K the proton–proton reaction dominates but at $\sim 1.7 \times 10^7$ K the CNO reaction reaches equality. For stars considerably more massive than

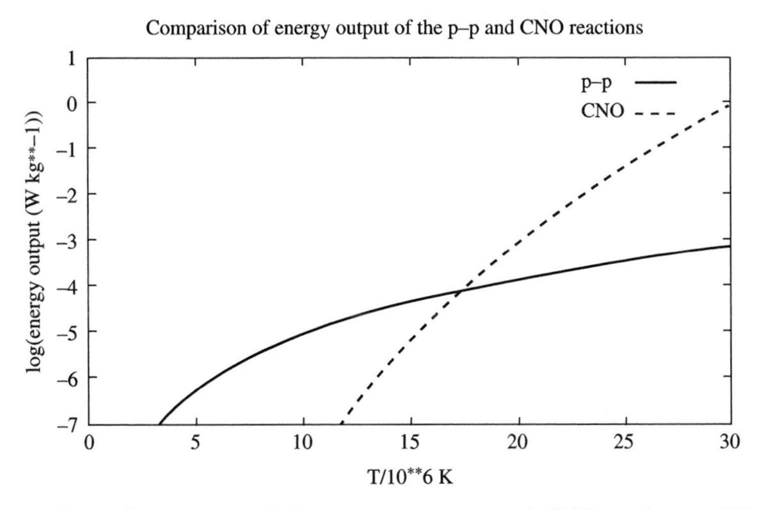

Figure C.4. Comparison of the proton–proton and CNO cycles at different temperatures.

the Sun, where higher temperatures prevail, the CNO reactions would certainly dominate.

Problems C

C.1 Protons of energy 10^{-16} J meet a rectangular potential barrier with a height of 10^{-15} J. If 10^{-30} of the impinging protons penetrate the barrier then what is its width?
(The mass of a proton is 1.673×10^{-27} kg)

C.2 The energy generation rate for the proton–proton reaction is $\varepsilon_{\text{pp}} = 9.5 \times 10^{-37} X_p^2 \rho T^4$ and that for the CNO cycle is $\varepsilon_{\text{CNO}} = 2.3 \times 10^{-8} X_p X_C \rho (T/10^7)^{17}$. Given the fractional content of hydrogen, X_p, is 0.7 and that of carbon, X_C, is 0.0003 then find the temperature at which the CNO generation rate is **(i)** one quarter of, **(ii)** equal to and **(iii)** four times that of the proton–proton cycle.

By what percentage is the total output of energy increased if the temperature is raised from 1.50×10^7 K to 1.51×10^7 K?

Appendix D

Radiation Pressure

Radiation within a black-body enclosure exerts a pressure on the enclosure walls. We can best imagine how this occurs by thinking of the electromagnetic energy in terms of photons, particles of energy $h\nu$ and momentum magnitude $h\nu/c$ where ν is frequency and c is the speed of light. Here we will first find the pressure in terms of the radiation energy density within the enclosure and then use the Planck radiation formula to find the energy density in terms of the temperature.

D.1. A Photon Model for Finding Radiation Pressure

The treatment we use here is a simplified one but it gives the correct answer. Consider a cubical box of side L, with internal reflecting walls, containing N photons of mean energy $h\bar{\nu}$ and mean magnitude of momentum $h\bar{\nu}/c$. The average momentum magnitude in each of the $x-$, $y-$ and $z-$directions is $h\bar{\nu}/(\sqrt{3}c)$ and the average speed in each of the directions is $c/\sqrt{3}$. Now consider a photon bouncing to-and-fro between the two walls perpendicular to the $x-$direction with those characteristics. At each bounce, since it reverses its direction of motion, there is a change of momentum of the photon with magnitude $2h\bar{\nu}/(\sqrt{3}c)$ and the time between such bounces is

$2\sqrt{3}L/c$. Hence the rate of change of momentum of the photon is

$$\frac{dp}{dt} = \frac{2h\bar{\nu}}{\sqrt{3}\,c} \div \frac{2\sqrt{3}L}{c} = \frac{h\bar{\nu}}{3L}. \tag{D.1}$$

This is the force on the wall of the enclosure; with N photons and wall area L^2 this gives the pressure on the wall as

$$P_{\rm rad} - \frac{Nh\bar{\nu}}{3L^3}.$$

The energy density within the enclosure, u, is $Nh\bar{\nu}/L^3$ so that

$$P_{\rm rad} = u/3. \tag{D.2}$$

The pressure is one third of the energy density for relativistic particles such as photons; in the case of gas pressure the factor is two-thirds.

D.2. The Energy Density from the Planck Radiation Equation

From equation (A.14) the energy density within a black-body enclosure for radiation with frequencies between ν and $\nu + d\nu$ is

$$E(\nu)d\nu = \frac{8\pi h\nu^3}{c^3} \frac{1}{\exp(h\nu/kT) - 1}\, d\nu. \tag{D.3}$$

The total energy density comes from

$$u = \int_0^\infty E(\nu)d\nu.$$

Changing the variable to $x = h\nu/kT$ we find

$$u = \frac{8\pi}{c^3h^3}\,(kT)^4 \int_0^\infty \frac{x^3}{e^x - 1}\, dx. \tag{D.4}$$

The integral in equation (D.4) is a standard integral and equal to $\pi^4/15$ giving

$$u = \frac{8\pi^5 k^4}{15c^3h^3}T^4 = aT^4, \tag{D.5}$$

where a is the *radiation constant.* It is related to Stefan's constant by $a = 4\sigma/c$. Combining this with equation (D.2) we have

$$P_{\text{rad}} = \frac{4\sigma T^4}{3c}. \tag{D.6}$$

Problem D

D.1 Find the temperature at which radiation pressure equals 10^5 Pa (Approximately atmospheric pressure).

Appendix E

Electron Degeneracy Pressure

Here we will use a simplified approach to derive an expression for electron degeneracy pressure. It will differ from that given in equation (4.9) by a numerical factor but will give the correct dependence on the properties of the material. The basic relationship on which the derivation depends is the *Heisenberg Uncertainty Principle*, expressed in the form

$$\Delta x \Delta p \geq h/2, \tag{E.1}$$

where, in one dimension, Δx is an uncertainty in position and Δp an uncertainty in momentum.

E.1. Position–Momentum Space

To completely define the state of a particle it is necessary to know both its position, defined by three coordinates in a Cartesian system, for example, and its momentum, defined by three components. Thus we can think of a six-dimensional space within which the particle exists. If we determine the x component of position and the uncertainty of position is Δx then it may be in the range $x - \Delta x$ to $x + \Delta x$. With this interpretation of uncertainty we may consider that for each particle there exists a six-dimensional volume, at least $\hbar^3$, within which the particle may be present. In a more precise formulation of the condition one would need to take into account that for both in position and momentum there is a probability distribution and the uncertainty is the standard deviation of a distribution. If the particles are fermions then they must satisfy the Pauli Exclusion Principle so that all particles within a particular system must be in different states. The following derivation assumes that each particle occupies a cell of volume $\hbar^3$, the minimum possible, in the position momentum space and that each such volume may contain two particles with opposite spins, corresponding to different states.

E.2. The Energy Density in Degenerate Material

We take degenerate material contained within a sphere of radius R. The number of particles contained within that volume with momenta between p and $p + dp$ is

$$dN = \frac{\frac{4\pi R^3}{3} \times 4\pi p^2 dp}{\hbar^3} \times 2, \tag{E.2}$$

where the final factor of 2 allows for particles of opposite spin to occupy the same cell. Hence the total number of particles within a sphere of radius R and with momenta with magnitudes between 0 and $p_{\max}$ is

$$N = \frac{32\pi^2 R^3}{3\hbar^3} \int_0^{p_{\max}} p^2 dp = \frac{32\pi^2 R^3 p_{\max}^3}{9\hbar^3}. \tag{E.3}$$

In the non-relativistic case the kinetic energy of a particle with momentum magnitude p is $p^2/2m$ where m is the particle mass. Hence the total kinetic energy for particles with momentum magnitudes between p and $p + dp$ is

$$dE_K = \frac{p^2}{2m} dN = \frac{16\pi^2 R^3 p^4}{3\hbar^3 m} dp,$$

so that the total kinetic energy is

$$E_K = \frac{16\pi^2 R^3}{3\hbar^3 m} \int_0^{p_{\max}} p^4 dp = \frac{16\pi^2 R^3 p_{\max}^5}{15\hbar^3 m}. \tag{E.4}$$

The term involving $p_{\rm max}$ can be eliminated from equation (E.4) by using equation (E.3). This gives

$$p_{\rm max} = \left(\frac{9\hbar^3 N}{32\pi^2 R^3}\right)^{1/3}, \qquad \text{(E.5)}$$

which gives

$$E_K = \frac{3}{10}\left(\frac{9}{32}\right)^{2/3} \frac{\hbar^2}{\pi^{4/3}} \frac{N^{5/3}}{R^3 m}. \qquad \text{(E.6)}$$

Now we imagine that we apply an external pressure very slightly greater than the degeneracy pressure and change the radius of the material sphere by an amount ΔR. Work will be done on the system and the kinetic energy will change by an amount ΔE_K. We can express this by

$$\Delta E_K = -4\pi R^2 P_d \Delta R,$$

where P_d is the degeneracy pressure. This gives, taking the limit $\Delta R \to 0$

$$P_e = -\frac{1}{4\pi R^2}\frac{dE_K}{dR} = \frac{9}{40}\left(\frac{9}{32}\right)^{\frac{2}{3}} \frac{\hbar^2}{\pi^{\frac{7}{3}}} \frac{N^{\frac{5}{3}}}{R^5 m}. \qquad \text{(E.7)}$$

The factor m in the divisor of equation (E.7) indicates that the main source of degeneracy pressure is electrons so we ignore the effect of nucleons and write m as the electron mass m_e and express the pressure as the electron degeneracy pressure, $P_{\rm ed}$. Now we transform (E.7) in a number of ways. Writing $4\pi R^3/3 = V$, the volume of the material and the ratio of the number of electrons to the number of nucleons, i.e. $r = N/N_{\rm nuc}$ we have

$$P_{\rm ed} = \frac{3}{40}\left(\frac{3}{8}\right)^{2/3} \left(\frac{rN_{\rm nuc}}{V}\right)^{5.3} \frac{\hbar^2}{m_e}.$$

Finally, to compare with equation (4.9) we introduce m_{nuc}, the average mass of a nucleon and write $\hbar = h/2\pi$ giving

$$P_{\text{ed}} = \frac{3}{40}\left(\frac{3}{8}\right)^{2/3}\frac{1}{\pi}\left(\frac{r\rho}{m_{\text{nuc}}}\right)^{5/3}\frac{h^2}{m_e}. \tag{E.8}$$

The numerical factor is about one-quarter of that in equation (4.9) but the dependency of the material properties is reproduced correctly.

Appendix F

The Eddington Accretion Mechanism

The Eddington accretion mechanism is described in terms of a simple model — a spherical body in a uniform stream of matter moving at some constant speed. Under suitable circumstances the body can accrete all the matter that falls upon it. The oncoming matter must impinge on the body with greater than the escape speed from the body. If the excess over the escape speed is sufficiently smaller than the bombarding matter shares its energy with surface material of the body and all the involved material then has less than escape speed and so is retained, i.e. *accretion* is taking place. However, if the arriving material is very much greater than the escape speed then, after the sharing of energy, both the oncoming matter and some of the surface material may have enough energy to escape so that *abrasion* of the body occurs. Here we are concerned with the case of total accretion.

F.1. The Accretion Cross Section

Figure F.1 depicts a spherical body of mass M and radius R situated in a uniform stream of material moving at speed V relative to the body. The figure shows various streams of matter which are focussed by the gravitational field of the body. The limiting stream, that marked OP, is that for which the matter arrives at the surface

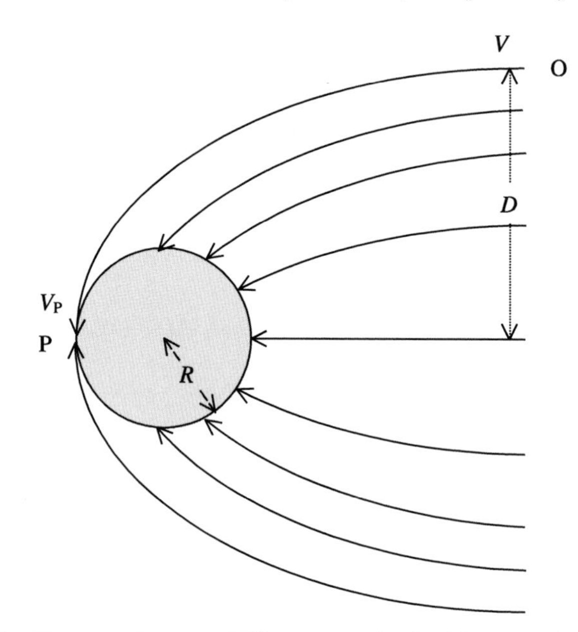

Figure F.1. Streams of matter falling on to a body. The accretion radius is D.

tangentially at P with speed V_P. The distance D is the *accretion radius* and the accretion cross-section, $A = \pi D^2$.

From conservation of angular momentum

$$VD = V_P R,$$

or

$$V_P = \frac{D}{R} V. \tag{F.1}$$

From conservation of energy

$$\frac{1}{2} V_P^2 = \frac{1}{2} V^2 + \frac{GM}{R}, \tag{F.2}$$

or, substituting for V_P from (F.1) and rearranging,

$$D^2 = R\left(R + \frac{2GM}{V^2}\right). \tag{F.3}$$

If V is very small then the second term in the bracket will dominate. If V is very large then the second term may be ignored and the accretion cross section is πR^2. In this case the flow lines are little affected by the body's gravitational field and are virtually straight.

The rate at which matter accreted is

$$\frac{dm}{dt} = \pi D^2 \rho = \pi R\left(R + \frac{2GM}{V^2}\right) V\rho, \qquad \text{(F.4)}$$

where ρ is the density of the medium within which the body is moving.

Problems F

F.1 Find the ratio D/R for $V = kV_{\text{esc}}$ where V_{esc} is the escape speed from an accreting body and $k = 0.1, 0.5, 1.0, 2.0, 5.0$ and 10.0.

F.2 A body with the mass and radius of the Sun is moving at a speed of $20\,\text{km s}^{-1}$ in a stationary medium of density $10^{-11}\,\text{kg m}^{-3}$. By how much does the mass of the body increase in 10^6 years?

Solutions to Problems

Chapter 1

1.1 The masses of oxygen per 100 gm of nitrogen are in the ratio 1:2:4 so without extra information the compounds could be NO, NO_2 and NO_4. However, this would give a nitrogen atom more massive than an oxygen atom. With the extra information other interpretations are possible with the ratios of nitrogen to oxygen m:1, m:2 and m:4, with m ≥ 2. The actual compounds are N_2O (nitrous oxide), NO (nitric oxide) and NO_2 (nitrogen dioxide).

The ratio of the mass of an oxygen atom to that of nitrogen is 57/50, or 1.14:1, which is close to the true ratio of 16:14.

1.2 **(i)** The speed of the electrons is

$$v = \frac{E}{B} = \frac{1{,}530}{1.332 \times 10^{-4}} = 1.149 \times 10^{7}\,\mathrm{m\,s^{-1}}.$$

(ii) The value of s, the deflection of the beam within the electric field region is 0.01 m, We have

$$\frac{e}{m} = \frac{2Es}{L^2B^2} = \frac{2 \times 1{,}530 \times 0.01}{(0.1 \times 1.332 \times 10^{-4})^2} = 1.725\times 10^{11}\,\mathrm{C\,kg^{-1}}.$$

(iii)

$$\frac{e}{m_0} = \frac{e}{m_c}\frac{m_v}{m_0}$$

$$= \frac{1.725 \times 10^{11}}{\sqrt{1-(0.1149/2.998)^2}} = 1.728 \times 10^{11}\,\mathrm{C\,kg^{-1}}.$$

1.3 With m the mass of the drop, q the charge on it, V the voltage across the plates and d their separation

$$mg = 6\pi\eta av = Vq/d,$$

and

$$6\pi\eta av = \frac{4}{3}\pi a^3\rho g \quad \text{giving}$$

$$a = \sqrt{\frac{9\eta v}{2\rho g}}$$

$$= \sqrt{\frac{9 \times 1.511 \times 10^{-5} \times 1.142 \times 10^{-4}}{2 \times 1{,}109 \times 9.8}} = 8{,}453 \times 10^{-7}\,\text{m}.$$

Hence

$$q = \frac{6\pi\acute{\eta}avd}{V}$$

$$= \frac{6\pi \times 1.511 \times 10^{-5} \times 8.453 \times 10^{-7} \times 1.142 \times 10^{-4} \times 0.004}{232.4}$$

$$= 4.732 \times 10^{-19}\,\text{C}.$$

This is nearly three times the magnitude of the electron charge (4.806×10^{-19} C).

1.4 The basic equation, from Einstein's equation linking mass and energy is

$$\frac{dm}{dt} = \frac{dE/dt}{c^2}.$$

Hence

$$\frac{dm}{dt} = \frac{3.83 \times 10^{26}}{(2.998 \times 10^8)^2} = 4.26 \times 10^9\,\text{kg}\,\text{s}^{-1}.$$

In 1 billion years the fraction of the Sun's mass lost will be

$$\frac{3.156 \times 10^{16} \times 4.26 \times 10^9}{1.989 \times 10^{30}} = 6.76 \times 10^{-5}.$$

Chapter 2

2.1 First we find the mass out to distance r. This is

$$M(r) = 4\pi a \int_0^r \left(1 - \frac{x}{R}\right) x^2 dx = 4\pi \alpha \left(\frac{r^3}{3} - \frac{r^4}{4R}\right).$$

Inserting this in (2.2) gives potential energy

$$\begin{aligned}
\Omega &= -16\pi^2 \mathrm{a}^2 \mathrm{G} \int_0^R \left(\frac{r^3}{3} - \frac{r^4}{4r}\right)\left(1 - \frac{r}{R}\right) r dr \\
&= -16\pi^2 a^2 G \int_0^R \left(\frac{r^4}{3} - \frac{r^5}{4R} - \frac{r^5}{3R} + \frac{r^6}{4r^2}\right) dr \\
&= -\frac{26}{315}\pi^2 a^2 G R^5.
\end{aligned}$$

Using $M(R) = \frac{\pi a}{3} R^3$ we find

$$\Omega = -\frac{26}{35} G \frac{M(R)^2}{R}.$$

2.2 From equation (2.5) the mass of the sphere is

$$M(R) = 2\pi \times 10^8 \times (10^8)^2 = 6.283 \times 10^{24}\,\text{kg}.$$

From (2.6) the gravitational potential energy of the sphere is

$$\Omega = -\frac{2}{3} \times 6.674 \times 10^{-11} \times \frac{\left(6.283 \times 10^{24}\right)^2}{10^8} = 1.756 \times 10^{31}\ \text{J}.$$

The number of molecules in the body is

$$N = \frac{6.283 \times 10^{24}}{4 \times 10^{-27}} = 1.571 \times 10^{51},$$

and if T is the average temperature then their average translational kinetic energy is $3kT/2$ Hence, from the Virial theorem

$$2 \times \frac{3}{2} kT \times N = -\Omega \quad \text{or}$$

$$T = -\frac{\Omega}{3kN} = \frac{1.756 \times 10^{31}}{3 \times 1.381 \times 10^{-23} \times 1.571 \times 10^{51}} = 270\,K.$$

2.3 **(a)** Starting with equation (2.16)

$$M_{Jc} = \left(\frac{375k^3}{4\alpha\mu^3 G^3}\right)^{1/2}\left(\frac{T^3}{\rho}\right)^{1/2} \quad \text{that can be rearranged as}$$

$$\mu = \left(\frac{375}{4\pi\rho M^2}\right)^{1/3}\frac{kT}{G}.$$

The density of the sphere is given by

$$\rho = \frac{3M}{4\pi R^3} = \frac{3 \times 2 \times 10^{30}}{4\pi \times (2 \times 10^{14})^3} = 5.968 \times 10^{-14}\,\text{kg}\,\text{m}^{-3}.$$

Hence

$$\mu = \left(\frac{375}{4\pi \times 5.968 \times 10^{-14} \times 4 \times 10^{60}}\right)^{\frac{1}{3}}\frac{1.381 \times 10^{-23} \times 20}{6.674 \times 10^{-11}}$$
$$= 2.069 \times 10^{-27}\,\text{kg}.$$

(b) First we calculate the geometrical moment inertia of a uniform sphere. The elements of mass are taken as shells of radius r and thickness dr all at the same distance, r, from the centre of the sphere. Hence

$$I = 4\pi\int_0^R \rho r^4 dr = \frac{4}{5}\pi\rho R^5.$$

Expressed in terms of the mass, $M = \frac{4}{3}\pi\rho R^3$

$$I = \frac{3}{5}MR^2.$$

With $\ddot{I}$ taken as constant

$$\dot{I} = 1{,}000t_y\ddot{I} = \frac{6}{5}MR\dot{R}, \quad \text{where } t_y \text{ is one year}$$

so that the surface speed of the protostar is

$$\dot{R} = \frac{5{,}000t_y}{6MR}\ddot{I} = \frac{5{,}000t_y}{6MR}(2K + \Omega).$$

The translational kinetic energy of the protostar material is

$$K = \frac{3}{2}kT\frac{M}{\mu}$$
$$= 1.5 \times 1.381 \times 10^{-23} \times 20 \sim \frac{2.4 \times 10^{30}}{2.069 \times 10^{-27}}$$
$$= 4.806 \times 10^{35}\,\mathrm{J}.$$

The gravitational potential energy is

$$\Omega = -\frac{3}{5}G\frac{M^2}{R}$$
$$= -0.6 \times 6.67410^{-11} \times \frac{(2.4 \times 10^{30})^2}{2 \times 10^{14}}$$
$$= -1.153 \times 10^{36}.$$

This gives

$$\dot{R} = \frac{5{,}000 \times 3.156 \times 10^7}{6 \times 2.4 \times 10^{30} \times 2 \times 10^{14}}$$
$$\times (2 \times 4.806 \times 10^{35} - 1.153 \times 10^{36})$$
$$= -10.5\,\mathrm{m\,s^{-1}}.$$

Note that the distance moved by the surface in 1,000 years is 1.66×10^{11} m, a tiny fraction of the original radius, which justifies the assumption that $\ddot{I}$ remains constant.

2.4 (a) The density of the sphere is 1.2 times the density found in 2.3(a) since the mass in 2.3(b) is greater by that factor and the radius is the same. Hence

$$\rho = 1.2 \times 5.968 \times 10^{-14} = 7.162 \times 10^{-14}\,\mathrm{kg\,m^{-3}}.$$

This gives the free-fall time as

$$t_{ff} = \left(\frac{3\pi}{32 \times 7.162 \times 10^{-14} \times 6.674 \times 10^{-11}}\right)^{1/2}$$
$$= 2.482 \times 10^{11}\,\mathrm{s} = 7{,}865\,\mathrm{years}.$$

(b) From (2.19)

$$\frac{dr}{dt} = -\left\{2 \times 6.674 \times 10^{-11} \times 2.4 \times 10^{30}\left(\frac{1}{1.5 \times 10^{14}} - \frac{1}{2 \times 10^{14}}\right)\right\}^{\frac{1}{2}}$$

$$= -731\,\mathrm{m\,s^{-1}}.$$

2.5 The length of a condensation is

$$l = \left(\frac{\pi kT}{G\rho\mu}\right)^{1/2}$$

$$= \left(\frac{\pi \times 1.381 \times 10^{-23} \times 30}{6.674 \times 10^{-11} \times 10^{-12} \times 4 \times 10^{-27}}\right)^{1/2}$$

$$= 6.982 \times 10^{13}\,\mathrm{m}.$$

The total mass of gas in a condensation, M, is $\pi r^2 \rho l$, where a is the radius of cross section of the filament. This gives

$$M = \pi \times 10^{26} \times 10^{-12} \times 6.982 \times 10^{13} = 2.193 \times 10^{28}\,\mathrm{kg}.$$

The Jeans critical mass is

$$M_{\mathrm{Jc}} = 5.46\left(\frac{k^3T^3}{\mu^3G^3\rho}\right)^{\frac{1}{2}}$$

$$= 5.46\left(\frac{(1.381 \times 10^{-23} \times 30)^3}{(4 \times 10^{-27} \times 6.674 \times 10^{-11})^3 \times 10^{-12}}\right)^{\frac{1}{2}}$$

$$= 3.338 \times 10^{29}.$$

The mass of a condensation in the filament has less mass than a Jeans critical mass so it will dissipate rather than collapse.

2.6 The Maxwell–Boltzmann distribution in terms of kinetic energy is of the form $f(K) = CK^{1/2}\exp\left(-\frac{K}{kT}\right)$ in which C

is a constant, so that

$$\frac{df(K)}{dK} = \frac{C}{2K^{1/2}}\exp\left(-\frac{K}{kT}\right) - \frac{CK^{1/2}}{kT}\exp\left(-\frac{K}{kT}\right).$$

This is zero when $K = \infty$, at the extreme tail of the distribution, and at $K = kT/2$, which is the peak value.

2.7 The cooling rates are:

T(K)	**Cooling rate *(10^5 W kg^{-1})**
50	2.34
150	4.60
250	4.55
350	4.27
450	4.00
550	3.75
650	3.54

These results, plotted as a smooth curve, are shown below.

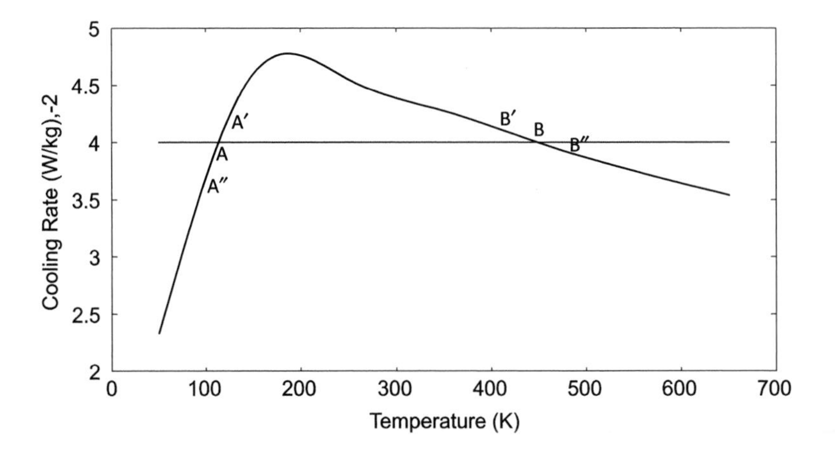

The horizontal line corresponds to the heating rate and equilibrium temperatures are at points A (~125 K) and B (~450 K). For point A, if there is a small perturbation taking the temperature to A′ then the cooling rate would increase and the temperature would fall and return to point A. If the perturbation was to A″ then the heating rate would exceed the cooling rate and, again, the temperature would return to point A. Hence, point A corresponds to stable equilibrium. For point B a small perturbation to B′ would increase the cooling rate and drive the temperature downwards, even further from B. Similarly, a perturbation to B″ would give a heating rate higher than the cooling rate, which would increase the temperature so moving the state even further from B. Hence, B represents a point of unstable equilibrium.

2.8 **(a)** The estimated opacity is $10^{-4}\,\mathrm{m^2\,kg^{-1}}$ so that the optical depth is $\tau = 10^{-4} \times 10^{-12} \times 20 \times 9.46 \times 10^{15} = 18.92$. Heat will leave slowly.

(b) The estimated opacity is $10^{-1}\,\mathrm{m^2\,kg^{-1}}$ so that the optical depth is $\tau = 10^{-1} \times 10^{-8} \times 10^{6} = 10^{-3}$. Heat will leave quickly.

2.9 From Wien's displacement law we have

$$T = \frac{2.898 \times 10^{-3}}{\lambda_{\mathrm{peak}}}.$$

For $\lambda_{\mathrm{peak}} = 4 \times 10^{-7}\,\mathrm{m}$, $T = 7{,}245\,\mathrm{K}$.
For $\lambda_{\mathrm{peak}} = 7 \times 10^{-7}\,\mathrm{m}$, $T = 4{,}140\,\mathrm{K}$.

2.10 From the Doppler equation

$$\begin{aligned} \mathrm{v} &= \mathrm{c}\frac{\mathrm{d}\lambda}{\lambda} \\ &= 2.99792 \times 10^{8} \times \frac{589.1053 - 588.9950}{588.9950} = 5.614 \times 10^{4}\,\mathrm{m\,s^{-1}}. \end{aligned}$$

Since this is positive it corresponds to a speed of $56.14\,\mathrm{km\,s^{-1}}$ away from the Earth.

Chapter 3

3.1 **(a)** The Jeans critical mass is

$$\begin{aligned} M_{\mathrm{Jc}} &= 2.0 \times 10^{21} \left(\frac{T^3}{\rho}\right)^{\frac{1}{2}} \\ &= 2.0 \times 10^{21} \left(\frac{500^3}{10^{-21}}\right)^{\frac{1}{2}} = 7.071 \times 10^{35}\,\mathrm{kg}. \end{aligned}$$

(b) From solution 2.3(b) the geometric moment of inertia

$$I = \frac{3}{5}MR^2.$$

With $\ddot{I}$ taken as constant and $t_y = 1$ year

$$\dot{I} = 10^6 t_y, \quad \ddot{I} = \frac{6}{5}MR\dot{R},$$

so that the surface speed of the protostar is

$$\dot{R} = \frac{5 \times 10^6 t_y}{6MR}, \quad \ddot{I} = \frac{5 \times 10^6 t_y}{6MR}(2K + \Omega).$$

The translational kinetic energy of the protostar material is

$$\begin{aligned} K &= \frac{3}{2}kT\frac{M}{\mu} \\ &= 1.5 \times 1.381 \times 10^{-23} \times 500 \sim \frac{5 \times 10^{35}}{4.0 \times 10^{-27}} \\ &= 1.295 \times 10^{42}\ \text{J}. \end{aligned}$$

The radius of the sphere is

$$R = \left(\frac{3M}{4\rho}\right)^{\frac{1}{3}} = \left(\frac{3 \times 5 \times 10^{35}}{4 \times 10^{-21}}\right)^{\frac{1}{3}} = 7.211 \times 10^{18}\,\text{m}.$$

The gravitational potential energy is

$$\begin{aligned} \Omega &= -\frac{3}{5}G\frac{M^2}{R} \\ &= -0.6 \times 6.67410^{-11} \times \frac{(5 \times 10^{35})^2}{7.211 \times 10^{18}} = -1.388 \times 10^{42}. \end{aligned}$$

This gives

$$\begin{aligned} \dot{R} &= \frac{5 \times 10^6 \times 3.156 \times 10^7}{6 \times 5 \times 10^{35} \times 7.211 \times 10^{18}} \\ &\quad \times (2 \times 1.295 \times 10^{42} - 1.388 \times 10^{42}) \\ &= 8.77\,\text{m}\,\text{s}^{-1}. \end{aligned}$$

Since the mass of the cloud is less that the Jeans critical mass then, as expected, the cloud is expanding.

Chapter 4

4.1 From equation (4.1)

$$R = \sqrt{\frac{\mathrm{L}}{4\pi\sigma\mathrm{T}^4}}$$

$$= \sqrt{\frac{3 \times 10^{30}}{4\pi \times 5.7 \times 10^{8} \times 3{,}900^4}} = 1.35 \times 10^{11}\,\mathrm{m} = 0.90\,\mathrm{au}.$$

The star is a red giant.

4.2 The mass of the ISM is

$$M_{\mathrm{ISM}} = \pi \times 10^{13} \div 4 \times (9.46 \times 10^{15})^3 \times 10^{-21} = 6.65 \times 10^{39}\,\mathrm{kg}.$$

The mass of stars is

$$M_{\mathrm{stars}} = 10^{11} \times 0.8 \times 1.989 \times 10^{30} = 1.59 \times 10^{41}\,\mathrm{kg}.$$

The proportion of ISM mass in the Milky Way is

$$P = \frac{M_{\mathrm{ISM}}}{M_{\mathrm{ISM}} + M_{\mathrm{stars}}} = \frac{6.65 \times 10^{39}}{6.65 \times 10^{39} + 1.59 \times 10^{41}} = 0.040.$$

4.3 Inserting values into equations (4.7), (4.8) and (4.9) give the following pressures in Pa:

	Gas	Radiation	Degeneracy
(i)	8.26×10^{10}	2.52	3.08×10^{13}
(ii)	8.26×10^{14}	2.52×10^{12}	1.43×10^{15}
(iii)	8.26×10^{15}	2.52×10^{16}	1.43×10^{15}
(iv)	8.26×10^{16}	2.52×10^{12}	3.08×10^{18}
(v)	8.26×10^{18}	2.52×10^{20}	3.08×10^{18}

Chapter 5

5.1 From equation (3.7)

$$P = \frac{10^4 \times 1.381 \times 10^{-23} \times 6 \times 10^6}{10^{-27}}\,\mathrm{Pa} = 8.286 \times 10^{14}\,\mathrm{Pa}.$$

For the small shift, $-\delta r$, towards the centre we assume that a finite-difference approach is justified so that, for example, from equation (5.1)

$$\begin{aligned}\delta P &= -\frac{GM(r)\rho}{r^2}\delta r\\ &= -\frac{6.674\times 10^{-11}\times 1.3\times 10^{30}\times 10^4}{(2.2\times 10^8)^2}\times(-10^7)\\ &= 1.793\times 10^{14}\,\text{Pa}.\end{aligned}$$

The positive sign indicates that pressure is higher and that the pressure is

$$P+\delta P = 1.008\times 10^{15}\,\text{Pa}.$$

Similarly, from equation (5.2)

$$\begin{aligned}\delta M(r) &= 4\pi r^2\rho\delta r\\ &= 4\pi(2.2\times 10^8)^2\times 10^4\times(-10^7) = -6.1\times 10^{28}\,\text{kg}.\end{aligned}$$

The included mass is lower and is 1.24×10^{30} kg.

From equation (5.7)

$$\begin{aligned}\delta T &= -\frac{3L\kappa\rho}{64\pi\sigma T^3 r^2}\delta r\\ &= -\frac{3\times 2\times 10^{26}\times 5\times 10^4}{64\pi\times 5.67\times 10^{-8}\times(6\times 10^6)^3\times(2.2\times 10^8)^2}\times(-10^7)\,\text{K}\\ &= 2.52\times 10^6\,\text{K}.\end{aligned}$$

The temperature is higher and is 8.52×10^6 K.

From equation (5.3)

$$\begin{aligned}\delta L &= 4\pi r^2\rho\varepsilon\delta r\\ &= 4\pi(2.2\times 10^8)^2\times 10^4\times 1.5\times 10^{-4}\times(-10^7)\\ &= -9.12\times 10^{24}\,\text{W}.\end{aligned}$$

The luminosity is lower and is 1.91×10^{26} W.

Chapter 6

6.1 The energy of the absorbed photon is $\Delta E = (10.38 - 5.74)e\text{V} = 4.64 \times 1.602 \times 10^{-19}$ J Hence, if the wavelength is λ then

$$\frac{hc}{\lambda} = 4.64 \times 1.602 \times 10^{-19}\,\text{J}.$$

Giving

$$\lambda = \frac{6.626 \times 10^{-34} \times 2.998 \times 10^{8}}{4.64 \times 1.602 \times 10^{-19}} = 2.67 \times 10^{-7}\,\text{m} = 267\,\text{nm}.$$

This is in the ultraviolet region of the electromagnetic spectrum.

6.2 The metallicity of the Sun is $1 - X - Y = 1 - 0.7381 - 0.2485 = 0.0134$.

One metallicity estimate for the star is $1 - 0.726 - 0.252 = 0.022$. For the other estimate we use the ratio of from to hydrogen. Thus we find

$$\begin{aligned}[\text{Fe/H}] &= \log\{(\text{Fe/H})_*/(\text{Fe/H})_\odot\} \\ &= \log\left\{\frac{0.00306/0.726}{0.0019/0.7381}\right\} = 0.211.\end{aligned}$$

Since this is positive the metallicity of the star is greater than that of the Sun and is $0.0134 \times 10^{0.211} = 0.022$. This is the same as the value of Z for the star. In general the two values are usually within 10% of each other.

Chapter 7

7.1 From equation (7.2) $D = 2/\beta = 2/3.72 \times 10^{-3} = 537.6$ pc. Percentage accuracy is $100 \times 0.02/3.72 = 0.54\%$.

Chapter 8

8.1 When the reading from B was made the nearby star was at a position such that, if a reading from A could have been made, it would have been at A'_2, the midpoint of A'_1 and A'_3. The angle between A_2' and B is $0.00524 + 1/2 \times 0.00316 = 0.00682''$. This corresponds to the angle β (Figure 7.7) in the case of a stationary star. Hence the distance of the star is

$$D = 2/0.00682 = 293 \text{ pc}.$$

The star moves through an angle $0.00316''$ in one year. In radians this angle is

$$\phi = \frac{0.00316}{3{,}600 \times 57.29} = 1.532 \times 10^{-8} \text{ radians}.$$

Hence the tangential speed of the star, where T_y represents a year, is

$$V = \frac{D\phi}{T_y} = \frac{293 \times 3.0857 \times 10^{13} \times 1.532 \times 10^{-8}}{3.156 \times 10^7} = 4.39\,\text{km}\,\text{s}^{-1}.$$

Its velocity is defined by the direction A_1' to A_3'.

The radial speed with respect to the Earth's position A_1 is

$$V_{A_1} = c\frac{d\lambda}{\lambda} = 2.998 \times 10^5 \times \frac{588.9791 - 588.9950}{588.9950} = -8.09\,\text{km}\,\text{s}^{-1}.$$

However, the radial speed with respect to the Sun = the radial speed of the star with respect to the Earth plus the radial speed of the Sun with respect to the Earth. The last speed is $+29.78\,\text{km}\,\text{s}^{-1}$. Hence the radial speed of the star with respect to the Sun is $-8.09 + 29.78 = 21.69\,\text{km}\,\text{s}^{-1}$, which is positive and hence away from the Sun.

Repeating the calculation for position B

$$V_B = 2.998 \times 10^5 \times \frac{589.0913 - 588.9950}{588.9950} = 49.02\,\text{km}\,\text{s}^{-1}.$$

The Earth at B is moving towards the star so the speed of the Sun with respect to the Earth is $-29.78\,\text{km}\,\text{s}^{-1}$. Hence the radial speed of the star with respect to the Sun is $49.02 - 29.78 = 19.24\,\text{km}\,\text{s}^{-1}$. The mean of the two estimates is $20.46\,\text{km}\,\text{s}^{-1}$.

Chapter 9

9.1 (a) Equation (9.7) is converted into the form

$$\log(d) = 1 - \frac{M - m}{5},$$

from which

$$\log(d) = 1 - \frac{5.7 - 19.1}{5} = 3.68.$$

Hence $d = 10^{3.68} = 4{,}790$ pc.

(b) From equation (9.7)

$$M = m + 5(1 - \log d) \quad \text{so that}$$

$$M = 6.4 + 5(1 - \log 153) = 0.48.$$

9.2 (a) From the converted equation in solution 9.1(a)

$$\log(d) = 1 - \frac{-2.03 - 20.4}{5} = 5.49.$$

Hence $d = 10^{5.49} = 30{,}900$ pc.

(b) If the apparent magnitude is 0 then

$$\log(d) = 1 - \frac{-2.03}{5} = 1.406.$$

Hence $d = 10^{1.406} = 25.5$ pc.

9.3 The distance is given by

$$\log(d) = 1 - \frac{-19.3 - 11.3}{5} = 7.12.$$

Hence $d = 10^{7.12} = 1.32 \times 10^7$ pc.

Chapter 10

10.1 The luminosity of the star is given by

$$\begin{aligned} L &= 4\pi d^2 b \\ &= 4\pi \times (310 \times 3.086 \times 10^{16})^2 \times 9.30 \times 10^{-10} \\ &= 1.07 \times 10^{30}\,\text{W}. \end{aligned}$$

The radius is given by

$$\mathrm{R} = \left(\frac{\mathrm{L}}{4\pi\sigma}\right)^{1/2}\frac{1}{\mathrm{T}^2}$$

$$= \left(\frac{1.07\times10^{30}}{4\pi\times5.67\times10^{-8}}\right)^{1/2}\frac{1}{3900^2} = 8.06\times10^{10}\,\mathrm{m}.$$

10.2 The radius is given by

$$r = \frac{9.05\times10^{16}}{M^{1/3}} = \frac{9.05\times10^{16}}{(1.6\times10^{30})^{1/3}}\,\mathrm{m} = 7\,740\,\mathrm{km}.$$

For a white dwarf $r \propto 1/M^{\frac{1}{3}}$ and for any body $M \propto r^3\rho$ Combining these gives $\rho \propto M^2$.

10.3 Since the neutron star has the same density as a neutron its radius, r_*, is related to the radius of the neutron, r_N, by

$$\left(\frac{r_\bullet}{r_N}\right)^3 = \frac{M_\bullet}{m_N} \quad \text{or} \quad r_\bullet = r_N\left(\frac{M_\bullet}{m_N}\right)^{1/3}.$$

Hence

$$r_\bullet = \left(\frac{4\times10^{30}}{1.67\times10^{-27}}\right)^{1/3}\times10^{-15}\ \mathrm{m} = \ 13.4\mathrm{km}.$$

For the period we have $r\cdot\omega^2 = 0.1\frac{GM_\bullet}{r_*^2}$ and $P = \frac{2\pi}{\omega}$. This gives

$$P = 2\pi\left(\frac{10r_\bullet^3}{GM_\bullet}\right)^{1/2}$$

$$= 2\pi\left(\frac{10\times(1.34\times10^4)^3}{6.67\times10^{-11}\times4\times10^{30}}\right)^{1/2} s = 1.9\,\mathrm{ms}.$$

Chapter 11

11.1 The sum of the semi-major axes of the orbits is, from (11.12),

$$a = (\alpha_1+\alpha_2)\frac{D}{2}$$

$$= \frac{0.00824\times318\times3.086\times10^{16}}{3600\times57.3\times2} = 1.96\times10^{11}\,\mathrm{m}.$$

The combined masses of the planets is given by (11.13) as

$$M = \frac{4\pi^2 a^3}{GP^2}$$
$$= \frac{4\pi^2(1.96 \times 10^{11})^3}{6.674 \times 10^{-11} \times (1.092 \times 3.156 \times 10^7)^2} = 3.75 \times 10^{30}\,\text{kg}.$$

The individual masses are given by a transformation of (11.14) as

$$M_1 = M\frac{\alpha_2}{\alpha_1 + \alpha_2} \quad \text{and} \quad M_2 = \frac{\alpha_1}{\alpha_1 + \alpha_2} \quad \text{giving}$$
$$M_1 = 3.75 \times 10^{30} \times \frac{0.00392}{0.00824}$$
$$= 1.78 \times 10^{30}\,\text{kg} \quad \text{and} \quad M_2 = 1.97 \times 10^{30}\,\text{kg}.$$

These are about 0.89 and 0.99 $M_\odot$ respectively.

11.2 From (11.16)

$$v_1 = 2.998 \times 10^8 \frac{0.0488}{2 \times 400.1664} = 18.28\,\text{km}\,\text{s}^{-1}$$

and

$$v_2 = 12.88\,\text{km}\,\text{s}^{-1}.$$

From (11.7)

$$a_1 = \frac{200 \times 24 \times 3600 \times 18280}{2\pi} = 5.03 \times 10^{10}\,\text{m},$$

and

$$a_2 = 3.54 \times 10^{10}\,\text{m} \quad \text{giving}$$
$$a = a_1 + a_2 = 8.57 \times 10^{10}\,\text{m}.$$

From (11.13) the combined mass is

$$M = \frac{4\pi^2(8.57 \times 10^{10})^3}{6.67 \times 10^{-11} \times (200 \times 24 \times 3600)^2} = 1.25 \times 10^{30}\,\text{kg}.$$

From (11.17)

$$M_1 = M\frac{v_2}{v_1 + v_2}$$

$$= 1.25 \times 10^{30} \frac{12.88}{18.28 + 12.88} = 5.17 \times 10^{29} = 0.26\, M_{\odot},$$

and

$$M_2 = 7.33 \times 10^{29}\,\text{kg} = 0.37\ M_{\odot}.$$

11.3 In the arbitrary units $b_1 + b_3 = 1$. When the smaller star is totally eclipsed the larger star provides all the brightness so that $b_1 = 0.933$ and $b_2 = 0.067$, in the transit situation

$$\left(1 - \frac{R_2^2}{R_1^2}\right) b_1 + b_2 = 0.945, \quad \text{or}$$

$$R_1 = R_2\left(1 - \frac{0.945 - b_2}{b_1}\right)^{-1/2} = 4.74\, R_{\phi}.$$

11.4 The following program in FORTRAN for finding the mass of the invisible member of an astrometric binary system is well annotated and can be written in other codes. To test it, put in the values for the mass of the visible star, radius of orbit and period given in the example in Section 11.6.3. For problem 11.4 the mass is 2.52 $M_{\odot}$.

```
PROGRAM ASTROMETRIC
C THIS CALCULATES THE MASS OF THE INVISIBLE STAR IN AN
C ASTROMETRIC BINARY

  DATA G,AU,YR,SM/6.673E-11,1.496E11,3.156E7,1.989E30/

C THE FOLLOWING GIVES THE MASS OF THE VISIBLE STAR (SOLAR
  UNITS),
C THE RADIUS OF ITS CIRCULAR ORBIT (au) AND THE PERIOD yr)

  DATA EM1,A1,P/1.15,18,70/

  EM1=EM1*SM
        A1=A1*AU
```

```
  P=P*YR
        PI=4.0*ATAN(1.0)

C CALCULATE COEFFICIENTS OF a^3 AND a^2

  C1=4*PI**2/G/P**2
  C2=C1*A1

C THE FOLLOWING CALCULATION FINDS f(a) FOR VALUES OF a FROM
C 10 au TO 1000 au AT INTERVALS OF 0.1 au. HOWEVER, IT STOPS
C WHEN TWO CONSECUTIVE VALUES HAVE OPPOSITE SIGNS AND USES
C LINEAR INTERPOLATION TO FIND THE VALUE OF A FOR WHICH f(A)=0

  V1=0
        V2=0
        DEL=0.1*AU
  DO 1 I=100,100000
  X=I*DEL
  F=C1*X**3-C2*X**2-EM1
        V1=F
        IF(I.EQ.100)GOTO 1
        IF(V1*V2.LT.0)GOTO 2
        V2=V1
  1 CONTINUE

C TWO CONSECUTIVE VALUES OF X ARE OF OPPOSITE SIGNS. LINEAR
C INTERPOLATION GIVES THE VALUE OF A FOR WHICH f(A)=0

  2 A=X-DEL*ABS(V1)/(ABS(V1)+ABS(V2))

C NOW THE VALUE OF A2 CAN BE FOUND

  A2=A-A1

C NOW THE VALUE OF EM2 IN SOLAR UNITS

  EM2=A1*EM1/A2/SM

WRITE(6,*)EM2
STOP
END
```

Chapter 12

12.1 (i) From equation (9.7) the absolute magnitude of 3C 273 is

$$M = 12.8 + 5\{1 - \log(7.49 \times 10^8)\} = -26.6.$$

(ii) Adapting equation (9.2) to absolute quantities the luminosity in solar units is

$$\frac{L}{L_\odot} = \frac{B}{B_\odot} = 10^{2(M_\odot - M)/5} = 10^{0.4(4.83+26.6)} = 3.7 \times 10^{12}.$$

This is an order of magnitude brighter than the Milky Way galaxy.

Chapter 13

13.1 From equation (13.2)

$$\begin{aligned} R &= \left(\frac{GM_* P^2}{4\pi^2}\right)^{\frac{1}{3}} \\ &= \left(\frac{6.674 \times 10^{-11} \times 1.2 \times 1.988 \times 10^{30} \times (4.62 \times 3.156 \times 10^7)^2}{4\pi^2}\right)^{\frac{1}{3}} \\ &= 4.410 \times 10^{11}\,\mathrm{m}. \end{aligned}$$

The speed of the planet in its orbit is

$$v_P = \omega R = \frac{2\pi R}{P} = \frac{2\pi \times 4.410 \times 10^{11}}{4.62 \times 3.156 \times 10^7} = 1.900 \times 10^4\,\mathrm{ms}^{-1}.$$

The minimum mass of the planet is

$$M_P = M_* \frac{v_*}{v_P} = 1.2 \times 1.988 \times 10^{30} \frac{9.0}{1.900 \times 10^4} = 1.13 \times 10^{27}\,\mathrm{kg}.$$

13.2 From equation (13.2)

$$\begin{aligned} R &= \left(\frac{\mathrm{GM}_* P^2}{4\pi^2}\right)^{\frac{1}{3}} \\ &= \left(\frac{6.674 \times 10^{-11} \times 1.3 \times 1.988 \times 10^{30} \times (2.83 \times 3.156 \times 10^7)^2}{4\pi^2}\right)^{\frac{1}{3}} \\ &= 3.266 \times 10^{11}\,\mathrm{m}. \end{aligned}$$

The speed of the planet in its orbit is

$$v_P = \omega R = \frac{2\pi R}{P} = \frac{2\pi \times 3.266 \times 10^{11}}{2.83 \times 3.156 \times 10^7} = 2.298 \times 10^4\,\mathrm{ms}^{-1}.$$

The mass of the planet is

$$M_P = M_* \frac{v_*}{v_P} = 1.3 \times 1.988 \times 10^{30} \frac{11.0}{2.298 \times 10^4} = 1.24 \times 10^{27}\,\mathrm{kg}.$$

The radius of the star is

$$r_* = 1.2 \times 6.955 \times 10^5 = 8.346 \times 10^5\,\mathrm{km}.$$

If the radius of the planet is r_P then from equation (13.5)

$$r_P = r_*(1 - 0.972)^{1/2} = 8.846 \times 10^5 \times \sqrt{0.028} = 1.48 \times 10^5\,\mathrm{km}.$$

13.3 There are many close commensurabilities

$$\frac{P_{\mathrm{Tethys}}}{P_{\mathrm{Mimas}}} = \frac{1.8878}{0.9424} = 2.003 \approx 2{:}1,$$

$$\frac{\mathrm{P}_{\mathrm{Dione}}}{\mathrm{P}_{\mathrm{Enceladus}}} = \frac{2.7369}{1.3702} = 1.997 \approx 2{:}1,$$

$$\frac{P_{\mathrm{Dione}}}{P_{\mathrm{Enceladus}}} = \frac{2.7369}{1.3702} = 1.997 \approx 2{:}1,$$

$$\frac{P_{\mathrm{Rhea}}}{P_{\mathrm{Dione}}} = \frac{4{,}5175}{2.7369} = 1.651 \approx 5{:}3,$$

$$\frac{P_{\mathrm{Hyperion}}}{P_{\mathrm{Titan}}} = \frac{21.2766}{15.9454} = 1.334 \approx 4{:}3.$$

These are the closest but there are some others fairly close — for example the ratio of the periods of Rhea and Enceladus is not far from 10:3.

Chapter 14

14.1 **(i)** The basic equation (14.12) is

$$\mathrm{r_c} = \left(\frac{2\pi \mathrm{D}_{\odot}^2}{\mu_0 \mathrm{v}\frac{\mathrm{dM}}{\mathrm{dt}}}\right)^{\frac{1}{4}}$$

$$= \left(\frac{2\pi \times 10^{50}}{1.2566 \times 10^{-6} \times 6 \times 10^5 \times 10^{13}}\right)^{\frac{1}{4}} = 3.021 \times 10^9\,\mathrm{m}.$$

This is 3.78 times the radius of the star.

(ii) First we must calculate the constant C from equation

$$\mathrm{C} = \frac{\beta r_{\mathrm{c}}^2}{\alpha R_*^2} - \frac{1+\alpha}{\alpha} = \frac{0.5 \times (3.021 \times 10^9)^2}{0.06(8 \times 10^8)^2} - \frac{1.06}{0.06} = 101.2.$$

The final spin rate is given by equation (14.18)

$$\Omega = \Omega_0 \left(\frac{M}{M_0}\right)^C$$

$$= 2.4 \times 10^{-4} \left(\frac{2.49}{2.5}\right)^{101.2} = 1.60 \times 10^{-4}\,\mathrm{radians\,s^{-1}}.$$

14.2 The Safronov equation to estimate formation time is

$$t_{\mathrm{form}} = \frac{3P}{\sigma(1+2\beta)} \left(\frac{4\rho_{\mathrm{sol}}}{3\pi^2}\right)^{2/3} M_p^{1/3}.$$

Inserting the values for Venus

(i)

$$t_{\mathrm{form}} = \frac{3 \times 0.615}{700 \times 9} \left(\frac{4 \times 4.4 \times 10^3}{3\pi^2}\right)^{\mathbf{2/3}} \times (4.868 \times 10^{24})^{1/\mathbf{3}}$$

$$= 3.51 \times 10^6 \text{ years}.$$

(ii)

$$\mathrm{t_{form}} = \frac{3 \times 29.46}{50 \times 9} \left(\frac{4 \times 4.1 \times 10^3}{3\pi^2}\right)^{2/3} \times (4 \times 10^{25})^{1/3}$$

$$= 4.53 \times 10^9 \text{ years}.$$

Chapter 15

15.1 Mean areal density is

$$\sigma_{\text{mean}} = \frac{0.01 \times 1.5 \times 10^{27}}{\pi \times (1.5 \times 1.496 \times 10^{11})^2} = 94.82\,\text{kg m}^{-2}.$$

15.2 Equation (15.13) is

$$t_{\text{sat}} = \frac{3P_s}{\sigma(! + 2\beta)} \left(\frac{4\rho_s}{3\pi^2}\right)^{\frac{2}{3}} M_s^{\frac{1}{3}}.$$

The period P_s is given by

$$\begin{aligned} P_s &= 2\pi\sqrt{\frac{R_s^3}{GM_P}} \\ &= 2\pi\sqrt{\frac{(4 \times 10^9)^3}{6.674 \propto 10^{-11} \times 2 \times 10^{27}}} \\ &= 4.351 \times 10^6\,\text{s} = 0.1379 \text{ years.} \end{aligned}$$

Hence

$$\begin{aligned} t_{\text{sat}} &= \frac{3 \times 0.1379}{4 \times 94.82 \times 9} \left(\frac{4 \times 3 \times 10^3}{3\pi^2}\right)^{\frac{2}{3}} (10^{23})^{\frac{1}{3}} \\ &= 3.08 \times 10^5 \text{ years.} \end{aligned}$$

Appendix A

A.1 The transformation is from $\langle E_\nu \rangle d\nu$ to $\langle E_\lambda \rangle d\lambda$ using

$$\nu = \frac{c}{\lambda} \quad \text{and} \quad d\nu = -\frac{c}{\lambda^2} d\lambda.$$

This gives

$$\langle E_\lambda \rangle = \frac{2hc^2}{\lambda^5} \frac{1}{\exp\left(\frac{hc}{\lambda kT}\right)}.$$

We simplify this to

$$\langle E_\lambda \rangle = \frac{A}{\lambda^5} \exp\left(-\frac{B}{\lambda T}\right).$$

Differentiating gives

$$\frac{d\langle E_\lambda \rangle}{d\lambda} = \frac{A}{\lambda^6} \exp\left(-\frac{B}{\lambda T}\right)\left(\frac{B}{\lambda T} - 5\right).$$

This is zero for $\lambda = \infty$ or $\lambda T = B/5$, corresponding to Wien's displacement law.

Appendix B

B.1 Transforming equation (B.8)

$$\beta = \frac{(1+z)^2 - 1}{(1+z)^2 + 1} = \frac{(1+0.4)^2 - 1}{(1+0.4)^2 + 1} = 0.324.$$

Hence the speed is $0.324c$.

Appendix C

C.1 Taking the natural logarithm of both sides of (C.2) and rearranging gives

$$d = \frac{1}{2} \ln\left(\frac{1}{P_{\text{pen}}}\right)\left(\frac{\hbar^2}{2m(\pounds_m - E)}\right).$$

For $P_{\text{pen}} = 10^{-30}$, $\hbar = h/2\pi = 1.055 \times 10^{-34}$ J, $m = 1.673 \times 10^{-27}$ kg, $E_m = 10^{-15}$ J and $E = 10^{-16}$ J we find

$$d = 2.10 \times 10^{-12}\,\text{m}.$$

C.2 $\frac{\varepsilon_{\text{CNO}}}{\varepsilon_{\text{pp}}} = \frac{2.3\times10^{-8}}{9.5\times10^{-9}} \frac{X_{\text{C}}}{X_{\text{p}}} \left(\frac{T}{10^7}\right)^{17} \frac{1}{T^4} = 1.04\times10^{-3} \left(\frac{T}{10^7}\right)^{13}$. If this ratio equals f then

$$T = 10^7 \left(\frac{f}{1.04 \times 10^{-3}}\right)^{1/13}.$$

This gives temperatures
(i) 1.52×10^7 K, **(ii)** 1.70×10^7 K, **(iii)** 1.89×10^7 K.
Energy generation at the two temperatures are as follows:

	1.50×10^7 K	1.51×10^7 K
p–p	$2.357 \times 10^{-8}\rho\,\mathrm{W\,kg^{-1}}$	$2.420 \times 10^{-8}\rho\,\mathrm{W\,kg^{-1}}$
CNO	$4.76 \times 10^{-9}\rho\,\mathrm{W\,kg^{-1}}$	$5.33 \times 10^{-9}\rho\,\mathrm{W\,kg^{-1}}$
Total	$2.833 \times 10^{-8}\rho\,\mathrm{W\,kg^{-1}}$	$2.953 \times 10^{-8}\rho\,\mathrm{W\,kg^{-1}}$

This corresponds to an increase of 4.2%.

Appendix D

D.1 Equating (D.6) to 10^5 Pa gives

$$T = \left(\frac{3 \times 10^5 c}{4\sigma}\right)^{1/4}$$

$$= \left(\frac{3 \times 10^5 \times 3.00 \times 10^8}{4 \times 5.67 \times 10^{-8}}\right)^{1/4} = 1.41 \times 10^5\,\mathrm{K}.$$

Appendix F

F.1 Since $V_{\mathrm{esc}}^2 = \frac{2GM}{R}$ we have

$$D^2 = R\left(R + \frac{2GM}{k^2 V_{\mathrm{esc}}^2}\right) = R\left(R + \frac{2GM}{k^2}\frac{R}{2GM}\right) = R^2\left(1 + \frac{1}{k^2}\right).$$

This gives $\frac{D}{R} = \sqrt{1 + \frac{1}{k^2}}$

V/V_{esc}	0.1	0.5	1.0	2.0	5.0	10.0
D/R	10.05	2.24	1.41	1.12	1.02	1.005

F.2 From (F.4) the mass gained in a time t is

$$\begin{aligned}\delta m &= \pi R\left(R + \frac{2GM}{V^2}\right)V\rho t \\ &= \pi \times 6.95 \times 10^8 \\ &\quad \times \left(6.95 \times 10^8 + \frac{2 \times 6.67 \times 10^{-11} \times 1.989 \times 10^{30}}{(2 \times 10^4)^2}\right) \\ &\quad \times 2 \times 10^4 \times 10^{-11} \times 3.156 \times 10^{13}\,\text{kg} = 9.15 \times 10^{27}\,\text{kg}.\end{aligned}$$

References

Anderson, D.R. *et al.* (2010), *Astrophys. J.*, **709**, 159–167.

Armitage, P.J. and Clarke, C.J. (1996), *Mon. Not. R. Astr. Soc.*, **280**, 458–468.

Bailey, M.E. (1983), *Mon. Not. R. Astr. Soc.,* **204**, 603–633.

Bailey, V. *et al.* (2014), *Astrophys. J. L.,* **780**, L4.

Beckwith, S.V.W. and Sargent, A. (1996), *Nature,* **383**, 139–144.

Benz, W., Slattery, W.L. and Cameron, A.G.W. (1986), *Icarus,* **66**, 515–535.

Bonnell, I.A., Bate, M.R. and Vine, S.G. (2003), *Mon. Not. R. Astr. Soc.*, **349**, 413–418.

Bonnell, I.A., Bate, M.R. and Zinnecker, R. (2005), *Proc. I.A.U. Symposium,* No. 227, eds. R. Cesaroni, M. Felli, E. Churchwell and C.M. Walmsley, Cambridge University Press: Cambridge.

Borucki, N.J. *et al.* (2011), *Astrophys. J.*, **736**, 19–40.

Butler, P. and Marcy, G. (1996), *Astrophys. J. L.*, **464**: L153.

Cameron, A.G.W. (1978), *The Origin of the Solar System,* ed. S.F. Dermott, Wiley: Chichester.

Chadwick, J. (1935), *Nobel Lecture: The Neutron and Its Properties,* Elsevier Publishing Company, Amsterdam. Available at https://www.nobelprize.org /uploads/2018/06/chadwick-lecture.pdf.

Chandrasekhar, S. (1931), *Astrophys. J.,* **74**, 81–82.

Clausius, R.J.E. (1870), *Philosophical Magazine.*, **40**, 122–127.

Cole, G.A.H. and Woolfson, M.M. (2013), *Planetary Science: The Science of Planets around Stars,* 2nd ed. p. 501, CRC Press: Boca Raton.

Cook, A.H. (1977), *Celestial Masers,* Cambridge University Press: Cambridge.

Dormand, J.R. and Woolfson, M.M. (1977), *Mon. Not. R. Astr. Soc.*, **180**, 243–279.

Fabrycky, D.C. and Tremaine, S. (2007), *Astrophys. J.*, **669**, 1298–1315.

Feuchtgruber, H. *et al.* (2013), arXiv:1301.5781[astro-ph.Ep].

Fowler, W.A., Caughlan, G.R. and Zimmerman, B. (1967), *Ann. Rev. Astron. Ap.*, **5**, 525–576.

Fowler, W.A., Caughlan, G.R. and Zimmerman, B. (1975), *Ann. Rev. Astron. Ap.*, **13**, 69–112.

Freeman, F.W. (1978), *The Origin of the Solar System*, Ed. S.F. Dermott, Wiley: Chichester.

Gaidos, E.J. (1995), *Icarus*, **114**, 258–268.

Gell-Mann, M. (1969), Nobel Lecture: *Symmetry and Currents in Particle Physics*, Elsevier Publishing Company, Amsterdam. Available at https://www.nobelprize.org/prizes/physics/1969/gell-mann/lecture.

Golanski, Y. and Woolfson, M.M. (2001), *Mon. Not. R. Astr. Soc.*, **320**, 1–11.

Goldreich, P. and Ward, W.R. (1973), *Astrophys. J.*, **183**, 24–26.

Gomes, R., Levison, H.F., Tsiganis, K. and Morbidelli, A. (2005), *Nature*, **435**, 466–469.

Goodricke, J. (1786), *Phil. Trans. R. Soc. Lond.*, **76**, 48–61.

Grossman, L. (1972), *Geochimica et Cosmochimica Acta,* **36**, 587–619.

Hayashi, C. (1966), *Ann. Rev. Astron. Astrophys*, **4**, 171–192.

Head, J.W. (1976), *Reviews of Geophysics and Space Physics*, **14**, 265–300.

Hewish, A., Bell, S.J., Pilkington, J.D.H., Scott, P.F. and Collins, R.A. (1968), *Nature*, **217**, 708–713.

Holden, P. and Woolfson, M.M. (1995), *Earth, Moon and Planets*, **69**, 201–236.

Holman, M.J. and Wiegert, P.A. (1999), *Astron. J.*, **117**, 621–628.

Hoyle, F. (1960), *Q. J. R. Astr. Soc.*, **1**, 28–55.

Hubble, E.P. (1937), *The Observational Approach to Cosmology*. Clarendon Press: Oxford.

Indulekha, K. (2013), arXiv.org>astro-ph>arXiv:1304.1554.

Jeans, J.H. (1902), *Phil. Trans. R. Soc. Lond.*, A**199**, 1–53.

Jeans, J.H. (1917), *Mon. Not. R. Astro. Soc.*, **77**, 186–199.

Kozai, Y. (1962), *Astrophys. J.*, **67**, 591–598.

Kroupa, P. (2001), *Mon. Not. R. Astro. Soc.*, **323**, 231–246.

Lamy, P.L. and Burns, J.A. (1972), *A. J. Phys.*, **40**, 441–444.

Laplace, P.-S. (1796), *Exposition du système du monde*, Cercle-Sociale: Paris.

Leavitt, Henrietta S. and Pickering, Edward C. (1912), *Harvard College Observatory, Circular.*, **173**, 1–3.

Lellouch, E. *et al.* (2001), *A&A*, **670**, 610–622.

Lemaître, G. (1927), *Un Univers homogène de masse constante et de rayon croissa rendant compte de la vitesse radiale des nébuleuses extra-galactiques* Annales de la Société Scientifique de Bruxelles, **47**, 49.

Loinard, L. *et al.* (2001), *Astrophys. J.*, **552**, L163–166.

Loinard, L. *et al.* (2002), *Planet. Space Science*, **50**, 1205–1213.

Lubow, S.H. and Ida, S. (2011), *Planet Migration*, ed. S. Seager, pp. 347–371, University of Arizona Press: Tucson, AZ.

Lucas, P. and Roche, P. (2000), *Mon. Not. R. Astr. Soc.*, **314**, 858–864.

Lynden-Bell, D. and Pringle, J.E. (1974), *Mon. Not. R. Astr. Soc.*, **168**, 603–637.

Mayor, M. and Queloz, D. (1995), *Nature*, **378**, 355–359.

McCaughrean, M.J. and Stauffer, J. (1994), *Astro. J.*, **108**, 1382–1397.

McConnell, A.J. and Woolfson, M.M. (1983), *Mon. Not. R. Astro. Soc.*, **204**, 1221–1240.

McCord, T.B. (1966), *Astro. J.*, **71**, 585–590.

McLaughlin, D.B. (1924), *Astrophys. J.*, **60**, 22–31.
Melita, M.D. and Woolfson, M.M. (1996), *Mon. Not. R. Astro. Soc.*, **280**, 854–862.
Michelson, A.A. and Morley, E.W. (1887), *American Journal of Science*, **34**, 333–345.
Millikan, R.A. (1924), Nobel Lecture: *The electron and the light-quant from the experimental point of view.* Available at https://www.nobelprize.org/prizes/physics/1923/millikan/lecture.
Nutzman, P.A., Fabrycky, D.C. and Fortney, J.J. (2011), *Astrophys. J. L.*, **740**, 10.
Observatoire de Paris (2004), Available at http://obspm.fr/how-did-the-planet-in-the-gamma-cephei-binary.html.
Oxley, S. and Woolfson, M.M. (2003), *Mon. Not. R. Astr. Soc.*, **343**, 900–912.
Oxley, S. and Woolfson, M.M. (2004), *Mon. Not. R. Astr. Soc.*, **348**, 1135–1149.
Parise, B. *et al.* (2002), *Astron. & Astrophys.*, **393**, L49–53.
Pauli, W. (1946), Nobel Lecture: *Exclusion Principle and Quantum Mechanics.* Available at https://www.nobelprize.org/prizes/physics/1945/pauli/lecture/.
Perlmutter, S. (2011), Nobel Lecture: *Measuring the Acceleration of the Cosmic Expansion Using Supernovae.*
Protszkov, E.-M., Adama, F.C., Hartmann, L.W. and Tobin, J.J. (2009), *Astrophys. J.*, **697**, 1020–1032.
Rogers, T.M., Lin, D.N.C. and Lau, H.H.B. (2012), *Astrophys. J. Lett.*, **754**, L6–L10.
Rossiter, R.A. (1924), *Astrophys. J.*, **60**, 15–21.
Roueff, E. *et al.* (2000), *Astro. & Astrophys.*, **354**, L63–66.
Rutherford, E. (1908), Nobel Lecture: *The Chemical Nature of the Alpha Particles from Radioactive Substances.* Available at https://www.nobelprize.org/prizes/chemistry/1908/rutherford/lecture/.
Safronov, V.S. (1972), *Evolution of the Protoplanetary Cloud and Formation of the Earth and Planets* (Israel Program for Scientific Translation, Jerusalem).
Schmidt, O.J. (1944), *Comptes Rendus* (Doklady) *Academie des Sciences de l'URSS*, **49**, 229–233.
Seaton, M. (1955), *Ann. Astrophys.*, **18**, 188–205.
Stevenson, D.J. and Salpeter, E.E. (1976), *Jupiter*, ed. T. Gehrels, pp. 85–112, University of Arizona Press: Tucson, AZ.
Stewart, G.R. and Wetherill, G.W. (1988), *Icarus*, **74**, 542–553.
Sumi, T. *et al.* (2011), *Nature*, **473**, 349–352.
Thomson, J.J. (1906), Nobel Lecture: *Carriers of Negative Electricity.*
Tillotson, J.H. (1962), *Tec. Rep. General Atomic Report G-3216.*
Tully, R.B. and Fisher, J.R. (1977), *Astro. Astrophys.*, **54**, 661–673.
Turon, C. *et al.* (1995). *Astro. Astrophys.*, **304**. 82–93.
Weidenschilling, S.J., Donn, B. and Meakin, P. (1989), *The Formation and Evolution of Planetary Systems*, eds. H.A. Weaver and L. Danley, pp. 131–150, Cambridge University Press: Cambridge.

Williams, I.P. and Cremin, A.W. (1969), *Mon. Not. R. Astr. Soc.*, **144**, 359–373.
Wolszczan, A. and Frail, D. (1992), *Nature*, **355**, 145–147.
Wood, K., Wolff, M.J., Bjorkman, J.E. and Whitney, B. (2001), *Astropys. J.*, **564**, 889–892.
Woolfson, M.M. (1964), *Proc. R. Soc.*, A**282**, 485–507.
Woolfson, M.M. (1979), *Phil. Trans. R. Soc. Lond.*, A**291**, 219–252.
Woolfson, M.M. (1999), *Mon. Not. R. Astr. Soc.*, **304**, 195–198.
Woolfson, M.M. (2003), *Mon. Not. R. Astr. Soc.*, **340**, 43–51.
Woolfson, M.M. (2004a), *Mon. Not. R. Astr. Soc.*, **354**, 419–426.
Woolfson, M.M. (2004b), *Mon. Not. R. Astr. Soc.*, **354**, 1150–1156.
Woolfson, M.M. (2007), *Mon. Not. R. Astr. Soc.*, **376**, 1173–1181.
Woolfson, M.M. (2011), *On the Origin of Planets: By Means of Natural Simple. Processes.* (a) pp. 348–350, (b) pp. 229–232, (c) pp. 269–283, (d) pp. 440–443, Imperial College Press: London.
Woolfson, M.M. (2013a), *Mon. Not. R. Astr. Soc.*, **436**, 1492–1496.
Woolfson, M.M. (2013b), *Earth, Moon and Planets*, **111**, 1–14.
Woolfson, M.M. (2016), *Earth, Moon and Planets*, **117**, 77–91.
Zel'dovich, Ya. B. and Raiser, Yu. P. (1966), *Physics of Shock Waves and High-Temperature Hydrodynamic Phenomena.* Academic Press: New York.

Name Index

W

Y

Z

Subject Index